DIY Music and the Politics of Social Media

Alternate Takes: Critical Responses to Popular Music is a series that aims to examine popular music from critical perspectives that challenge the accepted ways of thinking about popular music in areas such as popular music history, popular music analysis, the music industry, and the popular music canon. The series ultimately aims to have readers listen to – and think about – popular music in new ways.

Series Editors: Matt Brennan and Simon Frith

Editorial Board: Daphne Brooks, Oliver Wang, Susan Fast, Ann Powers, Tracey
Thorn, Eric Weisbard, Sarah Hill, Marcus O'Dair

Other Volumes in the Series:
When Genres Collide by Matt Brennan
Nothing Has Been Done Before: Seeking the New in 21st-Century American Popular Music by Robert Loss
Annoying Music in Everyday Life by Felipe Trotta
DIY Music and the Politics of Social Media by Ellis Jones
A Musical History of Digital Startup Culture by Cherie Hu (forthcoming)
Live from the Other Side of Nowhere: Contemplating Musical Performance in an Age of Virtual Reality by Sam Cleeve (forthcoming)
Ranting and Raving: Dance Music as Everyday Culture by Tami Gadir (forthcoming)
National Phonography: Field Recording, Sound Archiving, and Producing the Nation in Music by Tom Western (forthcoming)

DIY Music and the Politics of Social Media

Ellis Jones

BLOOMSBURY ACADEMIC
NEW YORK • LONDON • OXFORD • NEW DELHI • SYDNEY

BLOOMSBURY ACADEMIC
Bloomsbury Publishing Inc
1385 Broadway, New York, NY 10018, USA
50 Bedford Square, London, WC1B 3DP, UK

BLOOMSBURY, BLOOMSBURY ACADEMIC and the Diana logo
are trademarks of Bloomsbury Publishing Plc

First published in the United States of America 2021

Copyright © Ellis Jones, 2021

Cover design by Louise Dugdale
Cover image © Warmworld / iStock

For legal purposes the Acknowledgements on p. vi constitute
an extension of this copyright page.

All rights reserved. No part of this publication may be reproduced or
transmitted in any form or by any means, electronic or mechanical, including
photocopying, recording, or any information storage or retrieval system,
without prior permission in writing from the publishers.

Bloomsbury Publishing Inc does not have any control over, or responsibility for,
any third-party websites referred to or in this book. All internet addresses given in this
book were correct at the time of going to press. The author and publisher regret any
inconvenience caused if addresses have changed or sites have ceased to exist,
but can accept no responsibility for any such changes.

A catalog record for this book is available from the Library of Congress.

Library of Congress Cataloging-in-Publication Data
Names: Jones, Ellis (Ellis Nathaniel) author.
Title: DIY music and the politics of social media / Ellis Jones.
Description: New York City : Bloomsbury Academic, 2021. | Includes
bibliographical references. | Summary: "An investigation into the
contemporary practices of DIY musicians on social media and DIY's status
as "cultural resistance."" – Provided by publisher.
Identifiers: LCCN 2020033589 | ISBN 9781501359644 (hardback) | ISBN
9781501359668 (pdf) | ISBN 9781501359651 (epub)
Subjects: LCSH: Popular music–Social aspects. | Music trade–Social
aspects. | Social media.
Classification: LCC ML3918.P67 J67 2021 | DDC 306.4/842–dc23
LC record available at https://lccn.loc.gov/2020033589

ISBN:	HB:	978-1-5013-5964-4
	PB:	978-1-5013-5963-7
	ePDF:	978-1-5013-5966-8
	eBook:	978-1-5013-5965-1

Typeset by Integra Software Services Pvt. Ltd.

Series: Alternate Takes: Critical Responses to Popular Music

To find out more about our authors and books visit www.bloomsbury.com
and sign up for our newsletters.

CONTENTS

Acknowledgements vi

1 The problem?: Welcome to the democratic, DIY music business 1

2 The past: DIY, popular music and cultural resistance 23

3 The personal: Subjectivity and self-representation on social media 45

4 The players: Hierarchy, ownership and collectivism in DIY scenes 67

5 The public: Performing politics and elucidating difference 85

6 The popular: Metrics, measurements and the DIY imagination 103

7 The platform: Alterity and the political economy of social media 121

8 The plan … : Potential futures for DIY music, new media and social justice 137

References 150
Index 166

ACKNOWLEDGEMENTS

This book could not have existed without the enthusiastic cooperation of the DIY music practitioners in Leeds who gave their time to be interviewed and allowed me to pry into their emotions, experiences and practices. Although their words are presented anonymously here, they know who they are.

Most of the material here derives from my doctoral research at the School of Media and Communication, at the University of Leeds. My PhD supervisors there, Dave Hesmondhalgh and Leslie Meier, have both been incredible resources of knowledge, advice, support, critique and reassurance. I didn't realize how lucky I was at the time, but I'm starting to.

My PhD was funded, for the most part, by the UK's Arts & Humanities Research Council (grant number AH/L503848/1). Some of my doctoral research was published in fanzines – still available online – for which Emma Thacker and Daniel Howard provided beautiful illustrations. Some sections of this book have previously been published in *Popular Music and Society, Media, Culture, and Society,* and *Social Media and Society,* and are included here with permission.

Bethany Klein and Nick Prior offered essential comments (and confidence-boosting enthusiasm for this project) at my PhD examination in June 2018. Tami Gadir and Joey Whitfield gave important feedback at the book proposal stage. Matt Stahl and Kirsty Fife very kindly offered comments on my draft manuscript.

Ragnhild Brøvig-Hanssen has been very supportive of this project during a postdoctoral research period at the University of Oslo (enabled by the Research Council of Norway, project numbers 262762 and 275441). My time in Oslo would have been much worse without my beautiful boys: Emil Kraugerud, Tore Størvold and Kyle Devine.

Thanks are also due to the Sharrow Space Programme – especially Nick Clare for his tireless generosity – and to Patrick Hyland. It's a mark of his intelligence that he finds the academic world so baffling.

My family – my parents Helen and Barry Jones, my sister Libby Clark and her husband Bert, and my mum's mum, Bobba – have been ceaselessly encouraging throughout my education and into my scholarly 'career'.

Grace Whitfield has been invaluable from the first word to the last. Thank you.

1

The problem?

Welcome to the democratic, DIY music business

In the world of popular music, things – as always – aren't what they used to be. The once-rigid sureties of the music industries have been challenged by the disruptive potential of new technologies for making, circulating and experiencing music. We have arrived, apparently, at a radically different set of relations. As evidence, take this excerpt from the blurb of a recent book by Ari Herstand (2016), entitled *How to Make It in the New Music Business*: 'In the last decade, no industry has been through as much upheaval and turmoil as the music industry. If you're looking for quick fame and instant success, you're in the wrong field. It's now a democratic DIY business.'

The arrival of a democratic, DIY music business sounds like fantastic news, if true. It sounds like particularly fantastic news for any music practitioners who may have had an aversion to the 'old' music industries and who may have already been actively seeking and building alternative formations of popular music culture.

These alternative formations exist. Despite the above rhetoric of radical disruption, DIY music is not a new phenomenon. It is an approach to the production and distribution of music that, at a conservative estimate, dates back over forty years. It is often associated with punk, post-punk and indie, as well as electronic music genres, including rave. But more than an association with any particular genre, DIY scenes have historically often been affiliated with particular aspirations relating to the democratization of culture. They have questioned the organization and purpose of the music industries, and have called, implicitly and explicitly, for those industries to be radically re-organized or even wholly dismantled.

DIY might mean prizing the intimacy of a small venue, and the temporary community created within it, as an end in itself, rather than seeing it as a stepping stone. It might mean acknowledging the harmful aspects of competition invoked by a music industry that celebrates stars at the expense of valorizing a wider range of creative endeavours, and opting out of that race for fame and commercial success. It might mean seeing musical training as a manifestation of elitist distinction, and therefore emphasizing an 'anyone can do it' aesthetic over precise technical ability. These are some of the ways in which DIY music cultures have historically made claims for the distinctiveness of their approach in relation to the 'mainstream'.

But DIY today *is* mainstream. And when the case for the emergence of a 'DIY democratic music business' is made, the internet – and social media specifically – is usually offered as a major catalyst for such democratization. The internet, it is suggested, offers musicians a new, unguarded doorway to awaiting audiences. Media scholar David Croteau argues that 'while "independent," "alternative," and "DIY" media have long existed in many forms [...], one key to the Internet's unique significance is that it provides the infrastructure necessary to facilitate the distribution of all forms of self-produced media to a potentially far-flung audience' (2006: 341). Of course, the fact that this 'far-flung' distribution is *possible* does not mean that engagement with a worldwide audience is guaranteed, and it by no means assures the democratization of the media landscape, but it has certainly brought about substantial change.

Whilst social media may not be a panacea, what I wish to emphasize here is the extent to which it has *realized*, in a meaningful way, some of the core aspirations of DIY music, and has impacted on the lives of far more people than, say, punk ever did. Or to put it another way, there is a substantial overlap between the aspirations of DIY music and the kinds of communicative potentials opened up by social media. DIY has historically been presented as a story of people who *ought to be consumers* rejecting the role prescribed to them, turning the tables on 'popular culture' and becoming producers, and finding a sense of self-realization and political subversion in this act. Jello Biafra, singer of seminal US punk band The Dead Kennedys, has offered the mantra: 'don't hate the media, *become* the media' (Biafra 2000). This is, broadly, the promise of DIY, and it has also been a key promise of the internet and social media.

For better or worse, the lineage of DIY culture no longer has sole dominion over certain aspects of 'do-it-yourself' practice. DIY is increasingly acknowledged as an obvious choice for all sorts of musicians. Moving forward then, this book is premised on an understanding that there are two kinds of DIY music, which are at least theoretically separable. One is a broad but ultimately coherent tradition of cultural resistance, often undertaken in the name of greater aesthetic diversity, economic equality and access to participation – and often with inherent or implicit connections to larger ideas of social justice. The other is largely a socio-economic consequence of changes in the music industries, as well as in the ICT (information and communication technologies) industries. These changes in turn articulate to an increasingly prominent neoliberal discourse which emphasizes the need for individuals to 'take responsibility', rather than to seek or expect support from state or corporate institutions. What follows is an investigation into how these two versions of DIY music are interacting, and what the consequences are. If social media was the key tool by which popular music activity became increasingly 'DIY', what might it offer for music that was *already* DIY?

In defence of the alternative

This book is part of a series called Alternate Takes, which encourages its authors to challenge or re-frame conventional wisdoms in the world of popular music studies. When I proposed this book, my 'alternate take' was that, despite the rhetoric of democratization outlined above, social media has in lots of ways been quite bad for DIY music – at least, for the kind with a long history of politicized independence from the music industries. I still think this, and it is a key argument of the book. But this position feels far less controversial now than it did when I began my research in 2014. We are increasingly aware that the current, platform-dominated internet constitutes an extremely lopsided economy that is bad for musicians of all kinds and a communicative environment that, more generally, seems to be quite bad for all kinds of people.

Critical internet and social media scholars have problematized optimistic rhetorics of user empowerment and unfettered cultural production. They have highlighted the uneven economic relationship between a handful of platforms and their billions of users (McChesney 2013, Nieborg and Helmond 2018, Srnicek 2017a); suggested that new opportunities for autonomy (i.e. the freedom to act on one's own will, rather than following the dictates of others) might also lead to insecurity, compulsion and self-blaming (Duffy 2017, Kuehn and Corrigan 2013); and that the collection and application of data from our everyday online communication might represent the 'capture' of hitherto un-commodified dimensions of human activity (Andrejevic 2007, Dean 2010, Manzerolle and McGuigan 2014). As well as all this, the peak participatory 'moment' seems to be more or less over; platforms like YouTube increasingly play host to content produced by powerful 'old media' corporations (i.e. major labels, large film studios,

TV networks etc.), influenced by advertisers who 'do not want their advertisement next to low-quality home video content' (Kim 2012: 54).

What now feels more like the 'alternate take' is the idea that this politicized version of DIY music is something that is worth defending and protecting. There seems to be very little faith in 'alternative' music as a viable political project, and widespread scepticism that it even exists as something meaningfully distinct from other kinds of engagement with music. This scepticism is not new: the politicized distinction between 'mainstream' and 'alternative' music cultures is, in a sense, always facing an existential crisis. In the next chapter I suggest that this is a built-in consequence of DIY's ambivalent (i.e. love–hate) relationship to popular music. But it does seem that in the last two decades in particular, alongside the rise of social media and the new 'DIY' music business, a number of discursive threads have cumulatively questioned the idea that such claims to alterity could reflect anything other than a kind of social posturing.

There is a pessimistic, Frankfurt School-esque bent to this relativism: the idea that cultural choice is an illusion (Horkheimer and Adorno 2002), mangled as an apolitical postmodern cynicism. But its closer academic relative is Pierre Bourdieu's theory of cultural taste as 'distinction' (1984). It suggests that the only reason people are interested in 'indie' or 'underground' cultures is because it gives them a certain kind of credibility or status.

This intersects with a populist discourse – sometimes called 'poptimism' in music criticism circles (Rosen 2006) – which has questioned the political worth of any allegiance to alternative styles or scenes. This discourse negates a long-acknowledged tension between art and commerce by suggesting that popular music's commercial impetus is, in a sense, the very thing that forces it to engage with and reflect the cultural and political zeitgeist. Alternative music's relative hermeticism is consequently a source of aesthetic and political impoverishment. This in turn implies a kind of organic, frictionless inevitability to the social positioning of musical genres and traditions: alternative music is all fine and good for its own niche audience, it suggests, and mainstream popular music is good for its big, global audience.

Another discursive threat to DIY's validity relates to what the late cultural theorist Mark Fisher termed 'capitalist realism', whereby we come to see the presence of competitive market dynamics in our social lives as inevitable and unchangeable (2009). This is most evident, in this context, as a depoliticizing of cultural actors' decisions to engage with industries and practices that might once have been considered a betrayal of shared principles – the old notion of 'selling out' (Klein 2020; Klein, Meier and Powers 2017). These decisions are now seen as inviolably individual ('it's their choice') or as structurally overdetermined ('what else would you do?'), in such a way as to put them beyond critique. This perspective undermines any sense of collectivism in the setting of ethical norms and boundaries. The other consequence of this 'realistic' perspective is to see any alternative ethical position as, ultimately, a marker of privilege: for example, rejecting the profit motive is seen as a gesture only available to those with the economic security to afford it.

An increasing focus on representational politics in contemporary society – in both left-wing and right-wing forms – has devalued DIY's emphasis on organizational change as a mode of cultural resistance. DIY has historically valorized the building of alternative distribution networks, and these tend to come with inherent restrictions on audience size. But if the focus is on gaining representational visibility, then bigger is better. This perspective is reinforced as our understanding of the cultural industries – arriving both through academic research and through social media granting access 'behind the scenes' – draws new attention to the significant levels of agency operating within structures once caricatured as hegemonic monoliths (including major record labels). What makes the DIY musician so different, in terms of political potential, to the up-and-coming artist working to be heard in a fragmented and uncaring music industry? Isn't the latter navigating the same tensions between art and commerce, and perhaps negotiating them more successfully? The recent glut of best-selling, politically conscious, critically revered works from star US-based artists (King 2019) seems to beg the question of quite what the problem with the music industries was ever supposed to be.

I have sympathy for most of these arguments. I do think that music should be something it's possible to make a living doing, although I don't think that means accepting a moral equivalency between different kinds of music-making, or concluding that people should do 'whatever it takes' to make money. It's true that 'alternative' resentment towards chart music often continues a long history of misogynistic critique of young women's engagements with popular culture (Ewens 2019). It can also be dismissive or suspicious of African-American musics, sometimes seeing its capacity for 'technological innovation and stylistic change' as evidence of commercialism (Bannister 2006a: 88–9). And, undoubtedly, DIY does struggle to embrace and support the participatory diversity that is so central to its rhetoric. Some of this does relate to the uncommon material advantages that DIY practitioners might take for granted, although the economic security of DIY practitioners should not be assumed, and I think it's sometimes patronizing and wrong to suggest that not-for-profit principles inherently exclude certain social groups. (The tax-avoiding super-rich do not seem particularly interested in not-for-profit activity.)

The UK DIY scene that I have studied and been a part of seems at least as prone as other music scenes to abusive behaviour and prejudice. People I considered friends have taken advantage of their power, or of others' vulnerability, in a scene that was (and still is) specifically presented as a safer space. The question of whether DIY (in the specifically 'indie-punk' incarnation that I study here) is systemically sexist or racist is not one

I answer thoroughly here. But it is certainly true that it has often failed to properly account for intersectional injustices, tending to reflect instead the often-narrow social positions of its practitioners.[1] The scene I studied showed disheartening historical continuity in this regard: during my research period there were several flashpoints at which problematic racial politics were brought to the fore. There are times when DIY has been an important space for emancipatory struggles, most notably in its capacity to give voice to feminist and queer politics. But it's important to recall that this space has generally been hard-won by marginalized groups, rather than simply offered up willingly.

So, I will not at any point make the claim that DIY is an ideal kind of music culture. I don't think it constitutes anything so grand as a revolutionary political practice or a comprehensive social movement, and it also isn't unique in being a musical culture that shows marked differences to mainstream popular music, aesthetically, organizationally or economically. But I will claim that DIY, for all its imperfections, has the capacity to mitigate one problem in particular: the distance that popular music culture has from the lives of most people. Therefore, I offer a critical defence not of the entirety of DIY music as we find it now, but of the broader notion of the alternative – the idea that musical activity outside of the commercial popular music industries might bring us closer to experiences of culture that work towards and sometimes embody social justice. I suggest that DIY music has characteristics that can make it a valuable form of 'cultural resistance'. That's a term that has fallen out of favour somewhat in academic literature, and I attempt to justify my use of it in Chapter 2.

All of this doesn't say much about whether the music produced in DIY scenes is, in itself, at all superior or preferable to other kinds of music. That isn't really the focus of this book, which is more concerned with how DIY is organized, and how it communicates political values within and outside of its borders. DIY can sometimes be a space for music that seems to be commercially unviable, as in the kind of 'abrasive sonic tinkering' that Stephen Graham locates in his study of 'underground' music (2016:3); sometimes it is home to music that sounds quite similar to pop music found elsewhere. But regardless of the aesthetics that are favoured, I think the particular value of DIY is that it presents opportunities for a particularly close kind of 'articulation' (i.e. connection) between music and social life, which can (and sometimes does) have empowering, democratizing effects.

---

[1] Riot grrrl, a feminist, women-led DIY scene which began in the 1990s, is the DIY music lineage with the largest body of literature on experiences of exclusion in relation to race and ethnicity (see Bess 2015, Dawes 2013, Nguyen 2012). This is not to say that other scenes have not had comparable dynamics, but fewer accounts addressing them have been published.

DIY as the new default

So, DIY music is a cultural form with a long history of distinguishing itself from 'mainstream' music by means of specific ethical precepts. I've suggested that these might be valuable, and worth retaining and building upon. But social media and the internet have clouded some of the central ethical precepts of DIY music. These technologies intersect with, and often exacerbate, the existential crises I've listed above – of relativism, populism and pragmatism – as well as blurring distinctions between DIY and the music and ICT industries in other ways.

Take the 'not-for-profit' ethos as an example. DIY practitioners have historically tended to see a broad rejection of profiting from music, or variants on this theme (e.g. paying musicians but not promoters), as central to a vision of fair and ethical musical activity. But the internet has massively complicated notions of how much music costs, how much it *ought* to cost and even precisely what the music commodity is (Morris 2015). Automated surveillance of online activity is an important new site of profit which serves to underwrite 'free' access to culture in new ways (Andrejevic 2007), especially via targeted advertising, and many musicians today seek a similar kind of 'free lunch' model to that employed by tech companies. Street and Phillips, writing on music and copyright, quote one musician outlining such an approach: 'My attitude is like a start-up […] – you build up a community and then you monetize it […], give it away free, remove all the obstacles that would normally be there' (2016: 423). Clearly, this kind of approach to 'freeing' music is something quite distinct from a not-for-profit ethic. And there are concerns that the aspirational equation underpinning this activity – that free work now equals paid work later – might be economically infeasible and therefore subjectively harmful (Duffy 2017; Kuehn and Corrigan 2013).

Another key tenet of DIY has been independence or 'self-sufficiency'. This has been understood as important not only in terms of artistic autonomy, but also in order to have control over economic and organizational decisions that might otherwise exploit others (e.g. avoiding extortionate ticket prices). As noted above, the internet has been seen as a substantial boon for independent artists; even the internet itself has at times felt 'independent', insofar as its disruption of old music industry business models was sometimes presented as a grassroots, people-powered phenomenon. But the music industries being a more 'DIY business' means new expectations of 'self-management' – a form of independence that does not hold the same political potential. Record labels have become more risk-averse, and increasingly seek to shift the costs of production onto artists. Music industry scholars Mazierska, Gillon and Rigg suggest that major record labels now offer contracts 'only to those musicians who can prove their potential by having a significant following on social media or winning amateur competitions' (2018: 7). Part of DIY's approach, at least historically, has been to critique this notion of non-professional music as primarily a 'talent pool' for industry to draw from. That sense of DIY and mainstream music as 'separate worlds' can be hard to maintain today, when musicians of all kinds and all levels are using the same platforms.

DIY has often aimed to blur, or eradicate, distinctions between artists and audiences, aiming to increase participation by demystifying the practice of 'doing' popular music (either as a music performer or in other roles). To this end, they have used formats such as zines (i.e. small, hand-made or photocopied magazine-style publications) to encourage participation and have sought to build social and physical spaces that put musicians and audiences in close proximity. Social media does blur artist–audience boundaries; Nancy Baym has shown that 'getting closer' to audiences is a new requirement even for established, well-known musicians (2018). But this is a commercial imperative that relates to the commodification of our everyday communication (Dean 2010). Rather than demystifying cultural production, it can instead insert a new kind of mystification into the apparently 'direct' connection between fans and artists. Today, supposedly 'intimate' communications on social media can be very difficult to disentangle from the marketing and branding efforts that are part of so many popular musicians' diversified careers (Meier 2017). DIY practitioners consequently struggle to find an appropriately 'authentic' communicative mode by which they might avoid seeming (or feeling) grubbily self-promotional.

What I'm suggesting is that some of the ethical precepts that have long been central to DIY are also at least partly compatible with an emerging form of 'platform capitalism' (Srnicek 2017a). This could be an unhappy accident. But it is important to recognize that the capitalist class who, amongst other things, oversees the activity of the biggest music and ICT corporations, has a specific ability to absorb cultural critique and to work this critique into an augmented and thus re-legitimated economic system.

Luc Boltanski and Eve Chiapello's book *The New Spirit of Capitalism* (2005) represents an important attempt to theorize this relationship between capitalism and critique and, along with Hesmondhalgh and Meier's application of this material to the subject of independent music (2015), it is a key influence on the research I present here. Boltanski and Chiapello engage with Max Weber's concept of the 'spirit of capitalism' – that is, 'the ideology that justifies engagement in capitalism' (2005: 8) – in order to posit that this 'spirit' is often first presented, in a sense, by anti-capitalist critique. They show how the French radical politics of the 1960s offered up both an 'artistic critique' and a 'social critique' of labour conditions, and then outline how employers were able, in focusing on the artistic critique (which was about 'disenchantment' with a perceived spiritual paucity in everyday life), to instil work with new meanings that effectively dissolved the social critique (which had focused on inequality and suffering). Thus, flexibility that characterizes some kinds of work today is not an external arrival that was 'willed by no one' (2005: 185), but is in fact evidence of critique as a 'motor' of capitalist development (2005: 27).

So, economic and institutional power needn't always be employed as a top-down stifling of dissent (although this does happen); it can be involved in shaping more subtle processes of elision. Boltanski and Chiapello make a useful distinction between 'physical neutralization' (where critique 'does not succeed in being heard') and 'ideological neutralization' (where 'critique no longer knows what to say') (2005: 41). Symbols, texts, even whole domains of practice, can be 'hollowed out' and co-opted, whilst still carrying strong reverberations of their previous meanings. Since I'm suggesting that DIY still has the capacity to contribute to social justice, I'm also suggesting we need to be very careful of the consequences of these subtle elisions, and of the surface-level compatibility between DIY ethics and social media logics. And hopefully the overlaps outlined above also indicate why DIY music might be a specifically germane lens through which to consider social media's impact on society and culture more broadly.

As such, this book is just as much a study of social media as it is of popular music, but one that suggests social media research might fruitfully be carried out with a thorough understanding of the specific ethical precepts that people bring to social media platforms – in this case, the precepts of DIY music – and how they attempt to carry those ideas and behaviours with them when encountering these still new technologies. That doesn't mean that the normative perspective of the research has to come from the specific user-group under investigation. I have already suggested that I do have some affinity with DIY music's politics, but that isn't quite what I'm referring to here. I think that the ways that platforms constrain and enable behaviours can be best understood through a nuanced engagement with the social, political and cultural characteristics of a specific group of actually existing users. All of us arrive on social media with our own aims and intentions, however loosely or strictly defined, and with already-formed social groups that determine, at least in part, the social and cultural norms that will attempt to take root in those environments.

Nancy Baym, in her work on musicians, audiences and social media, uses the polysemy of the term 'platform' (i.e. its capacity to carry multiple meanings) as a starting point for considering links between online and offline communication environments, and the ways that both these spaces might carry some kind of 'architectural' power. 'Like concert halls', she writes, 'social media sites are built environments, designed to foster some social practices and discourage others' (2018: 155).

In this context, it is notable that DIY scenes have often placed emphasis on re-organizing spaces of musical activity. Some of the key calls to action in historical scenes have been broadly 'environmental' in this sense. Notable examples include the 'all ages' movement to allow under-21s into licensed venues in the United States in the 1980s, and riot grrrl's emphasis on 'girls to the front', which aimed to invert gendered audience norms in which women had often literally been at the periphery. In a slightly different, but related tradition of DIY music, we might also think about the environmental re-organization involved in the UK's outdoor rave scene of the 1990s, and how the movement of urban musics to rural settings related to desires for an anti-commercial musical space. In these ways and others, DIY scenes have highlighted the malleability of the relationship between built environments and the politics that operate in and around them. It's also important to note the sheer variety of platforms available online today, and the range of communication modes which even a single platform can offer. If you want to send private messages to a friend, or to video-chat with a group of colleagues, or to vent lengthy diatribes into the void, you can find and use online tools and platforms that are more likely to enable those kinds of communication.

So, we should be careful about assigning too much power to environments in themselves, whether mediated or physical. The issue, perhaps, is who has the power to alter them. As Baym notes, online 'built environments' are changing all the time. For better and for worse, Facebook and Twitter alter their design in response to how people are using them. I have written elsewhere that social media's 'affordances' – that is, the actions that they seem to enable and allow – should be considered critically as 'sites of contestation', where user intention and platform architecture meet (Jones 2019). In this book I try to focus on these moments and places where friction between the aims of users and platforms is evident, and to theorize outwards from there to reflect on broader tensions between DIY ethics and platform logics.

But is also important to understand that this 'architectural' approach to platforms only gets us so far. If we look at social media only as an environment that 'affords' certain things rather than others, then regardless of how we understand the power balance between users and platforms, we risk losing sight of social media as a broader societal force. Platforms have an influence on our life that goes beyond the observable decisions we make when interacting with and on the platform. Even your smarmy friend who is so quick to tell you that they've never had a Facebook account has nonetheless had their life substantially altered by the ways in which that platform has shaped contemporary life. Our relationships to friends and family, to local businesses, to politics, to culture and even to ourselves have all been changed, in indirect as well as direct ways. Therefore, social media's impact on a particular area of life – in this case, DIY music – may not be fully observable through studying usage in terms of the buttons that we do or don't press. It requires investigation into how social media practices fit, neatly or awkwardly, with the rest of life, and a consideration of online and offline activity as mutually constitutive.

Researching the Leeds DIY scene

Most of the material for this book comes from my study of a single DIY scene, undertaken in the city of Leeds between September 2015 and January 2018. Comparisons with other scenes and music cultures are made throughout. But a key premise, in the tradition of many other such studies of local music cultures (Cohen 1991, Finnegan 1988, Shank 1994), is that paying close attention to a single area of activity can reveal the operation of social mechanisms that would not be discoverable through wider-ranging surveys. The key advantage in this context is that studying a single scene, through a range of online and offline methods, permits a thorough assessment of the multiple roles that social media platforms play across different scales and kinds of communication. The findings outlined here relate not only to the ways in which the scene is outwardly communicative, but also to the 'everyday' individual and group interactions that construct and maintain it.

'Scene' is, of course, a term in popular usage, but it has also been employed as an academic concept in cultural studies (Straw 1990, 2001). Whilst the term can mask some quite different sociological approaches (Hesmondhalgh 2005: 28), there are some useful consistencies across its uses in popular music studies. Studying 'scenes' tends to involve paying particular attention to their 'overlapping' nature, and thus to a certain mobility of membership, whilst also pointing to the existence of something more stable than the posited hyper-flexibility of postmodern identity (i.e. the notion that we can 'make' ourselves into whoever we want to be). It also highlights that my interest in music-making here is not only in the final 'product', but in the routines and rituals that maintain and re-produce a music culture. Simon Frith notes that the concept of scene might usefully emphasize 'banality' whilst still celebrating 'some kind of opposition to dominant ideology' (2004: 174); Keith Kahn-Harris's account of extreme metal fandom similarly considers the way in which scenic 'mundanity' is a necessary ballast that allows for experiences of 'transgression' (2004). Even the most aesthetically uncompromising scenes tend to be characterized by repetition and stability: the same sets of people doing the roughly same things in roughly the same places.

Choosing Leeds was partly a matter of convenience, based on my location and my existing position in this community (outlined below). But Leeds also offered access to a wider range of people, activities and venues than in other nearby cities. Leeds also has an active and longstanding connection to DIY, most famously in its hosting of a highly political post-punk scene in the 1970s and 1980s (O'Brien 2012).

Speaking in terms of genre, I label my research population as broadly 'indie-punk' in a concession to its two clearest ideological lineages and to distinguish it from other local and trans-local DIY scenes centred around hardcore punk, electronic music, grime, folk and so on. There is a general tendency towards guitars and away from electronic instruments, and a construction of authenticity that tends to rely on some tropes drawn from rock music but which also reflects an increasing enthusiasm for popular music. So, while 'DIY' (rather than 'indie' or 'punk') is a label which captures this scene's valorization of particular methods of production and circulation, this doesn't mean genre is negated entirely. Indeed, we might even understand the term 'DIY' as problematic insofar as it makes an excessive claim to genre-indifference, which obscures the role of aesthetic discrimination in forming scene boundaries.[2]

Practitioners involved in this 'indie-punk' scene were mostly white and middle class, fairly mixed in terms of gender and sex (with a strong interest in feminist and queer politics), mostly vegetarian and vegan, politically left-leaning but not necessarily vocal or radical, and were mostly aged between eighteen and forty. In terms of social and cultural capital, then, there are commonalities that bind this scene together beyond generic affiliation.

However, above genre and status, I stress the role of place, and specifically venues. There are a number of venues that help constitute and maintain the scene, and one in particular serves to help define my research population. Wharf Chambers is a worker's cooperative and members' club with a bar and multi-use venue, which is open every day, and hosts several music events each week. Located in the city centre, near the so-called Freedom Quarter that denotes a cluster of LGBT-oriented venues, it emerged from a previous venue, Common Place, which was formed in the same location in 2005. Temple of Boom is another important city-centre venue (without a real 'bar' space outside of the gig room) which tends towards heavier punk and metal; Chunk is a practice space and gig venue in Meanwood operated by a collective of bands and artists, with an emphasis on art-rock and esoteric electronic music.

There are larger venues, too, which play a role in the scene's construction. Brudenell Social Club is a multi-room venue in Hyde Park which has received national recognition within the live music industry (Live Music Awards 2015), and which tends to host bigger indie, pop and rock acts in its 300 capacity main room. DIY music practitioners, however, had played a key role in its gradual transformation from working men's

---

[2] A sense of shared, politicized identity can also help practitioners to see beyond genre – in many DIY scenes it is primarily queer identity that plays this role. Pearce and Lohman's research on trans music scenes finds a similar kind of leniency towards genre in operation (2019).

club to student-facing venue, and so retained some sense of attachment, alongside some disappointment that the Brudenell had 'outgrown' the DIY scene. Local pubs like the Fox and Newt in Burley, and The Fenton and The Packhorse in the university area of Woodhouse, still hosted occasional shows. Belgrave Music Hall and Headrow House, two city-centre venues operated by one local company, overlapped with the DIY scene insofar as practitioners would attend (and play as opening acts at) bigger shows there but, for the most part, these two venues were seen to embody a different set of values, reflected in more self-conscious, faux-industrial interior design, as well as expensive beer and ticket prices.

Wharf Chambers in particular, though, was central. In particular, its status as a cooperatively run, queer-friendly venue, with a safer spaces policy, vegan food and relatively affordable prices, allowed it to stand in for and symbolize the values held by the scene. The understanding was, broadly: *if it happens at Wharf, it's DIY*. Even as different nights brought in overlapping but distinct crowds, the sense of a coherent scene hinged on a shared affinity with and attachment to place. This also demonstrates how local and trans-local notions of the DIY scene might relate – through similar experiences of attachment to DIY venues across the country (and beyond), members of the UK-wide scene felt as though they 'belonged' at Wharf, even though they may only visit once a year when touring. Indeed, Wharf often served as a model for those seeking a stable 'home' for DIY in their own city or town.

This broad veneration of Wharf Chambers gave way to a more complex relationship during the course of my research. A specific accusation of abuse against a Wharf staff member led to broader concerns about the venue's accountability processes, especially in relation to racism and discrimination. This led to, amongst other things, an informal boycott of Wharf by some members, the formation of an action group (Wharf Members against Racism) and the venue drawing up an anti-racism action plan (which, as of November 2019, readily admits 'previous failings'). During this period several members' meetings took place which I did not attend, since I felt my presence as a researcher might well inhibit participants' willingness to speak on these sensitive topics. So, whilst those events aren't fully documented here, a generalized concern with regards to the 'whiteness' of the scene provides an important context for this research on DIY's relationship to social media. Such concerns reinforce the need to engage seriously with the historical and contemporary dimensions of racism in DIY (as well as problematic histories regarding ableism and heterosexism), and particularly in punk rock (see Duncombe 2011). We certainly cannot take any association between DIY and social justice as inherent – it is for that reason that I attempt a critical engagement with the notion of 'cultural resistance' in Chapter 2.

In tracing this DIY scene online, I followed other digital culture researchers in thinking that the boundary of online study ought to reflect the usage pattern of the research population in question, where possible (Stirling 2016: 63). Nancy Baym uses the metaphor of the 'pub crawl' to consider how the most appropriate object of study is not one single online institution amongst many, but the meanings created by a set of actors who traverse across these spaces (2007).

The most commonly used site was Facebook and Facebook Pages (which has a standalone app but is within the Facebook ecosystem). All of my research participants had some degree of administrative control over a Facebook Page – for their band, solo music project, gig promotion, record label, venue, studio, practice space and in many cases several of the above – and the majority also maintained a personal Facebook profile. Twitter was the next most popular general-purpose social media platform, although usage here was more varied and several interviewees claimed to not really 'get' its purpose.

The most commonly used music-hosting site was Bandcamp. Bandcamp is a privately owned music hosting and sales platform founded in 2007, which fulfils digital music sales and mediates sales of physical goods, and almost all my participants had access to at least one artist or label page on the site. Whilst Bandcamp has links to Silicon Valley through its CEO and early investors, it has a reputation for being 'indie' and artist-friendly and is, unlike other comparable music streaming services, regularly turning a profit. SoundCloud offers similar services (although emphasizes streaming and embedding capabilities, rather than sales), but was used by fewer Leeds DIY practitioners, had more 'industry' associations and was considered primarily to be a home for electronic music genres.

Music streaming platforms, such as Spotify, iTunes (now Apple Music), Google Play and so on, were less central to the scene, since they generally do not allow the kind of free, instant account-creation and music-uploading that characterizes Bandcamp and SoundCloud. Rather, these platforms have aimed to get bigger labels and publishers on board in an attempt to create a music catalogue that will appeal to a broad consumer base (Hesmondhalgh, Jones and Rauh 2019); independent artists are required to go through third-party distributors (such as Record Union or Tunecore), most of which charge annual music registration fees, and to then wait for their music to be approved and uploaded. This is changing fast, as Spotify playlists become an increasingly powerful form of 'exposure', and the process of dealing with these third-party distributors becomes easier (i.e. more automated) and cheaper.

YouTube offers, like Bandcamp and SoundCloud, the ability to upload material quickly and without cost, and potentially to a far greater audience than these specialist independent music sites. It was generally used by practitioners for hosting music videos for 'singles' (i.e. lead tracks from releases), or other one-off videos, and wasn't home to much intra-scene communication. The notion of being a 'YouTuber' carried connotations of brand-building and self-absorption that my interviewees sometimes saw as contradictory to DIY ethics. YouTube's parent company, Alphabet, was part of the everyday online experience for practitioners in various forms, including email, file sharing, scheduling and in the prevalence of Google Search as a means of information retrieval. Of particular importance to the scene was the understanding of urban space enabled by Google Maps, and the associated information provided by Google Places.

Photo and video-centric platforms Instagram and Snapchat do not feature heavily here, since they were used by only a few practitioners, but they are becoming central for everyday communication and also gathering audiences for circulating music. Private chat applications like Facebook Messenger and WhatsApp were widely used but difficult to observe. Ticket sales sites, merchandize ordering and fulfilment sites, and file-hosting sites are also part of the online infrastructure that supports and shapes the scene.

Whilst we seem to be in a period of relative stability, there is no guarantee that the current key online platforms will stick around – a similar research project undertaken ten or fifteen years earlier may have noted the seemingly unbreakable dominance of MySpace and the prevalence of local music forums in organizing and maintaining scenes. Those platforms that do last tend to meddle with site architecture incessantly and also adapt their business models in order to keep up with competitors. I have tried to keep this in mind when writing, with the aim of sustaining the value of this research in the longer term by focusing not on specific platform functionalities, but on the relatively stable relationships that form between platforms' textual and architectural characteristics, and practitioners' aims and perspectives (Hesmondhalgh, Jones and Rauh 2019).

Interviews and observations are the two primary sources of data utilized in this project. I conducted twenty-four semi-structured interviews with twenty-eight Leeds-based DIY music practitioners[3] between August 2015 and August 2016. Material from these interviews is included throughout

---

[3] I use the term 'practitioner' throughout the book, rather than 'musician', because it is characteristic of DIY scenes that people hold more than one role. These practitioners included musicians, promoters, producers, sound engineers, visual artists and venue staff, with most holding at least two of these roles. Leeds-based music writer Rob Hayler, although writing about a slightly different, more experimental scene in the city, introduces the useful notion of the 'no-audience underground': 'There is no "audience" as such, in the sense of "passive receivers", because almost everyone with an interest in the scene is involved somehow in the scene. The roles one might have – musician, promoter, label "boss," distributor, writer, "critic," paying punter and so on – are fluid, non-hierarchical and can be exchanged or adopted as needed' (Hayler 2015).

Chapters 3–8, and all interviewees have been anonymized.[4] I attended a large number of gigs in Leeds during my research period, and was also given access to some rehearsals, recordings and meetings (although that 'behind the scenes' material did not prove hugely relevant). I also observed scene members' online practice, paying specific attention to prominent Leeds-specific Facebook Groups and Pages. This online observation fed back into offline interviews, where discussion would often turn to a specific event or interaction that had taken place online.

Alongside interviews and observations, the other key source of material has been my own involvement in DIY music. DIY has played a more formative role in my life than almost anything else I can think of. It is a place where I have learned about politics and ethics, founded and re-enforced numerous lasting friendships, and had my most profound experiences of music, provoking both personal reflection and collective exuberance. In Bristol, the city where I grew up, to discover a local musical world apart from the charmless, extortionate pubs we had been playing in as teenagers was to discover a culture that felt valuable and powerful in a way that nothing had previously, with connections to other local and national scenes that suggested a movement at once both globally visible and intimately secret. The Bristol scene had (and still has) a particularly strong identification with feminist and queer politics, as well as with veganism, and these particular integrations of political thought and action with musical culture rang true for me. They felt full and rich where previously posited connections with music and politics (in mainstream folk, punk, reggae and dance) had felt shallow. Whilst I would consider myself more open to other musical and political worlds now, and more aware of DIY's own particular foibles and flaws, the connection has nonetheless been a lasting one. Much of the last ten years has been spent, to the detriment of any other interests, playing in bands and putting on gigs, and meeting people with similar shared passions.

When my doctoral research brought me to Leeds in 2014, I co-founded a non-profit DIY promotion collective with the few friends I already half-knew. It was a fantastic way to divide up the sometimes-formidable labour of

---

[4]Julia Downes, Maddie Breeze and Naomi Griffin have written thought-provokingly on the specific ethical considerations of conducting research with DIY cultures. In their work, the issue of anonymizing or pseudonymizing data is examined as a power relation between researcher and participant, particularly on those occasions where participants might *want* to be named and recognized as 'critical agents of social change' rather than 'objects' to be observed (Downes, Breeze and Griffin 2013: 106–7). However, in this book I have opted to anonymize the DIY practitioners I spoke with, since much of the material contains opinions and perspectives on other local institutions and practitioners. This material is important to the research, but also has the potential to cause ill-feeling, and therefore I consider anonymity to be the best means of ensuring that the trust placed in me by those people is not used recklessly. Participants are instead numbered randomly (P1, P2, P3 etc.).

organizing shows – booking bands to play, promoting the show online and off, cooking dinner for the performers (and baking cakes for the audience), running the zine stall, occasionally doing the sound (badly) and providing somewhere for the bands to sleep. That collective lasted for two years, and there was some other DIY music activity too, playing music in my own band as well as other people's projects, and attending countless shows. For the final ten months or so of my research I was living in Sheffield – an hour's train journey from Leeds – and becoming involved in that city's DIY music scene, although in more peripheral roles.

As well as my participation in various DIY scenes, I also had a rather different set of engagements with music culture over the same period. During the same month I started my research, I signed a recording contract with an independent label who were, unbeknownst to me at the time, in partnership with Caroline International – a subsidiary of Universal Music Group, and very much part of the 'industry'. My musical venture, which sat somewhere between indie rock band and solo recording project, released two albums through that label in 2015, supported by frequent touring and other promotional activity. We had some press coverage in the sort of publications that even my parents had heard of – *Pitchfork, Rolling Stone, NME, Guardian* – and got to meet and play with musicians that remain heroes and role models to me (although I have local, DIY heroes too). Whilst the band and I found this to be a level of 'success' that both surprised and, at times, perturbed us, our record sales (and assorted income streams) weren't as strong as the label had anticipated, and the option to extend my contract wasn't taken up. Our most recent albums were self-released on cassette and vinyl, and all our touring returned to being self-organized.

Running alongside this academic research on DIY, then, was a strange parallel journey through the industry. Perhaps unsurprisingly, there were numerous times when it felt like the boundaries between research and practice were hard to define (as well as boundaries between work and leisure). This project then has been informed by my own experiences in ways that would be difficult to document fully. The imposition of third parties into our working practice as a band – PR companies, booking agents, tour managers – gave me an understanding of how artists' autonomy is 'negotiated' within music industry management structures (Banks 2007: 7). Touring with musicians (as well as meeting other industry workers) from the UK and the United States who manage to make a living from music gave me insight into their positive experiences as cultural workers, as well as the sacrifices involved, and the extent to which self-management has become a defining characteristic of work in this specific corner of the cultural industries (again, Banks 2007). And during this time, I found my relationship to the DIY scene felt increasingly problematic – my band was still referred to frequently in interviews and features as a

'DIY' project and yet we were really anything but, having 'sold out' at least by the standard measure of signing a record contract and taking accompanying steps towards professionalization. At times I felt partially responsible for (or at least compatible with) some of the harmful kinds of individualist aspiration which I identify and examine in this book. So, although this book offers an 'insider' perspective on DIY music, it also reflects my experience 'outside' of DIY – as musician and researcher – and hopefully this allows for sensible consideration of the value offered by DIY's specific political project.

This book is a work of social science, and it therefore carries the peculiar reflexivity characteristic of that discipline: in attempting to study some existing social phenomenon, it can change the object of study itself. That is a sufficiently daunting notion so as to inspire a cautious, considered approach towards research and the presentation of new knowledge. But, in evaluating the relationship between social media platforms and DIY music, I do hope that my research will have some impact on the ways in which practitioners engage with platforms, with each other and with the wider world. I think that DIY music continues to offer, at its best, a strong form of resistance to some forms of social injustice. The critical examination offered here is intended to bolster that strength. I follow Rebecca Solnit in thinking that 'authentic hope requires clarity' (2006: 20), and therefore this critical examination of DIY is not intended to be a fault-finding inquisition, but rather a consideration of the ways in which it is threatened by new forms of capitalist accumulation, put forward in the belief that these threats can and should be countered.

Chapter outlines

In this opening chapter, I have shown that a certain kind of 'DIY' activity is increasingly proposed as a means of successfully operating around and within the contemporary music and ICT industries. This discourse has much in common with a longer-standing conception of DIY music, especially in the emphasis placed on autonomy, participation and independence. But it also differs in some substantial ways from this older DIY lineage, which has often emphasized its distinction from and incompatibility with the 'mainstream' music industries on ethical and economic grounds. I have suggested that social media is a key site where the differences between these discourses of DIY might be elided, and where a previously 'resistant' cultural approach might be made compatible with increasingly normalized expectations of entrepreneurial self-government.

Chapter 2 engages with the history of DIY music, in order to give context to the present-day scene in Leeds and to further elucidate how and why DIY has been differentiated from other popular musics. It outlines DIY's historical relationships both to popular music and to communication technologies, drawing on examples from notable past DIY scenes – including UK post-punk, US indie and riot grrrl. It also engages with 'cultural resistance' – an often-maligned concept that I argue retains some power to explain DIY's specific political orientation.

The rest of the book is structured around findings drawn from fieldwork and theoretical interpretations of those findings. The chapters are ordered so as to approximate an 'outward' progression in terms of the social units considered, starting from the personal dimensions of online communication and moving towards 'larger' objects of study: the DIY scene; its relationship to local publics and other scenes; its relationship to commercial popular musics; and finally its relationship to the economic specificities of online platforms.[5]

So, Chapter 3 deals with DIY practitioners' approach to personal communication and identity construction online. It highlights that the communicative intimacy that was previously a distinctive characteristic of DIY is today increasingly compelled as part of 'doing' social media. Social media's tendency towards 'self-branding' and 'relatability' is partly mitigated by DIY's aversion to commercialization, but nonetheless means practitioners are faced with a complex set of norms to navigate. Affective engagement at this individual level is characterized by feelings of social anxiety.

Chapter 4 considers the role of social media in mediating relationships within the DIY scene. Offline, DIY scenes do have hierarchies and gatekeepers, but there also tends to be a high degree of collective ownership over various organizations. Online, a brief moment of devolved ownership (forums and stand-alone sites) was replaced with a move to Facebook and the like. Forms of collectivism demonstrated offline can prove difficult to apply within the frameworks of major platforms. In general, network structures serve to reinforce Romantic (i.e. anti-social) notions of creativity and authorship.

Chapter 5 considers 'the public' – that is, how DIY music might relate to the 'outside world' and particularly to other local music scenes. This is an area in which the DIY scene remains quite distinct. A discourse of 'safe spaces' is central to the scene's efforts to create an insular, protective environment and social media's 'echo chamber' can assist in extending this approach online. The consequences of this insularity are considered, both as

---

[5]This book structure owes a substantial debt to David Hesmondhalgh's *Why Music Matters* (2013).

valuable in some cases and as potentially restrictive in others. I also show, through a case study of an argument on social media, how interactions with other music scenes are shaped by platform design as well as by user intention.

Chapter 6 considers the role that online platforms play in mediating the relationship between DIY scenes and commercial popular music. Platforms place DIY musicians in the same space as global pop superstars, and metrics (the quantitative measurements of Followers, Friends and Likes that are abundant on social media) offer new capacities of direct comparison between these entities. In showing this vast gulf in scale, social media mitigates the capacity of DIY practitioners to imagine their scene as 'alternative' in a sense other than 'niche'. Nonetheless, metrics do continue to provide some ambivalent value by re-presenting local, material practice in a way that offers a sense of security and self-affirmation.

Chapter 7 outlines DIY's relationship to the platform economy and expands on the themes covered in this introductory chapter. DIY and other music scenes are increasingly making use of the same online platforms and digital tools. Whilst this brings new opportunities for self-organization and creative autonomy, these strategies of resourcefulness are also in keeping with an individualized and neoliberal 'enterprise discourse', wherein doing it 'yourself' loses much of its radical alterity. I also demonstrate, using Harry Braverman's work in the field of labour process theory, that reliance on automated tools provided by monopolistic platforms constitutes a relative 'deskilling' of DIY culture. I introduce 'optimization' as an important platform logic: a means by which consumer autonomy is employed to justify the use of marketing strategies by producers.

Finally, in Chapter 8 I offer some thoughts on how DIY music scenes might engage with social media platforms in order to best enable cultural resistance. I suggest that this might include utilizing free and/or open-source software, building cooperative platforms and networks, and moving away from future-oriented brand-building.

SUGGESTED LISTENING

All the following chapters are interspersed with 'suggested listening' boxes, which act a bit like epigrams. They put forward songs that, during the writing of this book, seemed to be speaking – abstractly or concretely – about themes I was trying to engage with. They aren't DIY songs necessarily, and they aren't all by artists whose politics I would align with. I could have quoted the lyrics in the text, but that would defy the point, which is that music is a distinct form of communication which relates to academic research in strange ways that can feel at once tangential and vital. Paul Simon's 'When Numbers Get Serious' (1983) is a goofy reggae song with an even goofier triple-time ending, and a silly, dad-joke lyric full of arithmetic wordplay. But then he sings: 'I will love you innumerably/you can count on my word,' and it's heart-breaking, and suddenly I feel like I understand more about the ethical distinctions between qualitative and quantitative epistemologies. I've made Spotify and YouTube playlists compiling the songs mentioned here, which you can find by searching for the title of this book, but both of these are incomplete (and likely impermanent) in different ways because of rights restrictions on those platforms. It has been argued that this kind of fragmentation of the media landscape motivates users to illegally access music (Bode 2019); maybe this fragmentation could also encourage other kinds of ethical engagements with music consumption.

2

The past

DIY, popular music and cultural resistance

The term 'DIY' has been used to describe a wide variety of musical activity and also to link musical activity to a wide variety of social and political aspirations. Since one of the primary aims of this book is to assess change – namely, how has social media changed DIY? – it is necessary to do some conceptual pruning in order to isolate DIY music as a meaningful social abstraction, whilst also acknowledging there is no single, definitive form of DIY music. That 'pruning' is the main aim of this chapter, and it is accordingly structured as a journey away from generality and towards specificity.

The first section offers a very broad look at DIY activity and DIY culture, as well as at the diversity of DIY music scenes operating in genres and in parts of the world that are not a core consideration in the rest of the book. This is not an ecumenical effort to bridge differences, but rather a deliberate attempt at being 'over-inclusive' in order to think sensibly about scale and scope: how 'big' is DIY music, in the context of other political and social entities? What non-musical activity can it be meaningfully connected to? What areas of people's lives does it touch upon? To what extent do global DIY scenes present a compatible notion of 'alternative' music?

In the second section, I narrow the focus further in order to consider the specific lineages of DIY most relevant to the Leeds scene that is the central focus of this book. Even in the relatively coherent 'indie-punk' tradition of DIY, different historical scenes have had quite different aims, and quite different aesthetic and communicative styles. I offer a brief overview of three such scenes – UK post-punk, US post-hardcore indie and riot grrrl – and establish them as important points of comparison for the chapters that follow.

In the third section, drawing on the material discussed in the first two sections, I argue that DIY music's relationship to popular music is *ambivalent*. DIY music is often in opposition to popular music, but it is also enamoured with the things that popular music can do, and generally seeks to emulate (and augment) the main forms and practices of popular music. I argue that this matters, since studies of DIY music have often overstated the distinctiveness of DIY, presenting it as a coherent social movement or as radical political praxis. I suggest that this ambivalence creates a particular set of tensions that are relatively consistent across diverse DIY scenes, even as their political goals and musical styles might differ.

I conclude by arguing that DIY can be meaningfully considered as 'cultural resistance'. However, this term is often used uncritically and thus can oversimplify the relationship between DIY and 'mainstream' culture. It is therefore valuable to establish a normative framework that is distinct from DIY practitioners' own perspectives on what constitutes cultural resistance. A broad notion of social justice, drawing on the political philosophy of Nancy Fraser, might be a starting point for that normative assessment.

A wide world of DIY activity

The connotations of the term 'DIY' extend beyond music and beyond the realm of cultural activity. This much is evidenced by my editor's insistence that this book's title ought to contain the word 'music', in order to avoid attracting an unwarranted audience in search of home improvement advice. It's true that I will not be offering any tips on wallpapering or furniture assembly here (and I am very, very far from being an expert on such subjects), but it is worth pointing out that the connection between these two kinds of DIY is not just a linguistic accident. Some of the meanings they carry are related: to engage with any kind of DIY activity is inherently to engage with the distinct values and meanings that we ascribe to production and consumption, that is, what it means to 'do', rather than to have others do for us; what kind of financial or ethical value is ascribed to these activities; how such activity ought to be divided up in terms of class, gender, age and so on. Central to most conceptions of DIY, musical or otherwise, is the notion that production offers some specific ethical benefits that consumption does not – either for the individuals doing the producing or via its wider social ramifications.

On a very large scale, DIY activity can constitute an intervention at the level of social infrastructure. This is the kind of activity documented in Kimberley Kinder's book, *DIY Detroit* (2016). During her fieldwork in 2012–13, Kinder found that parts of this predominantly African-American city were so devoid of adequate public services that residents undertook various forms of 'urban self-provisioning' in the 'gray spaces' that 'private owners abandoned and public officials neglected'. This included maintaining community gardens, street cleaning and acting as informal 'real estate agents' for dilapidated properties in their area. Kinder does not romanticize this activity, arguing that only a lucky few were able to treat it as 'an opportunity for countercultural experimentation'. For most it was a necessity, 'since existing market practices and government policies did not meet their basic needs' (Kinder 2016: 6–9). The feminist political theory of Gibson-Graham has highlighted similar kinds of 'DIY' activity (although perhaps in less desperate conditions) and has focused on the affective dimensions of these kinds of alternative economies. They argue that in building small anti-capitalist spaces, practitioners reshape understandings of what 'the economy' is, countering its reification by 'cultivating ourselves as subjects who can imagine and enact a new economic politics' (Gibson-Graham 2006: xxviii).

♪ 'Miracles will start to happen' – **Jonathan Richman**

Such examples could be considered a 'social' kind of DIY rather than a 'cultural' (or musical) kind, since they are not especially concerned with the 'symbolic creativity' that we might say defines cultural activity (Hesmondhalgh 2019: 9–10). But sometimes these two dimensions are understood as connected, as they are in George McKay's book *DiY Culture: Party and Protest in Nineties Britain* (1998). He identifies a kind of 'cultural turn' in political activism, away from labour and trade union politics and towards a 'youth-centred and directed cluster of interests and practices around green radicalism, direct action politics, [and] new musical sounds and experiences' (1998: 2). He is particularly focused on UK-based 'anti-roads' protests and their socio-political connection to electronic dance music culture of the period, which was specifically criminalized by the Conservative government's infamous 1994 Criminal Justice and Public Order Act outlawing 'music [...] characterised by the emission of a succession of repetitive beats'. McKay argues that there is a deep coherence here between musical and political community and suggests that 'anything from a drop-in advice centre to a living space is evidence of DiY's aim to combine party and protest, to blur the distinction between action and living' (1998: 26).

A sociological research project undertaken in and around Bristol in the 1990s takes a similar approach in describing an 'extended milieu' of DIY cultures and considers music festivals, veg boxes and LETS (local-exchange trading schemes) to all be part of this same research object (Purdue et al. 1997). I think anyone reading from 'my' scene would be highly sceptical of the idea that paying for delivery of organic vegetables constitutes DIY activity (although clearly it may once have done, to some people). The point worth retaining, though, is that such 'milieus' can be sociologically meaningful, even if we might like to think we each have our own highly distinct set of pursuits and interests. One of the interview participants in this Bristol-based study, when told that it focused on veg boxes, local trading exchanges and music festivals, responded: 'So, you're researching my whole life then!' (Purdue et al. 1997: 648).

Clearly, musical DIY cultures can have a perceived ethical continuity with other aspects of life that might also be considered as alternative or resistant. However, I think it's useful to set limits on this continuity. DIY music is meaningfully different from something like a 'social movement' – even though there are bound to be overlaps between music scenes and political activities. As the second half of this chapter outlines, DIY music simply doesn't have (or want) the distance from popular music that would make it a cultural equivalent either to living 'off-grid' or to plotting a complete political revolution.

Another aspect of DIY's conceptual breadth is the diversity of music scenes that have been identified by scholars as being DIY, or which have self-defined as such. DIY music activity is taking place in a wide range of locations, as well as across many musical genres. Such diversity raises the question of whether all such activity is best considered as belonging to the same category of musical culture.

For example, Elham Golpushnezhad's study of underground rap music in Iran outlines a DIY scene that is in many respects strikingly distinct (2018). One key distinction is that the pathway towards professional work in the music industries is practically non-existent, at least in the rap genre. So, while we might usually define DIY by its oppositionality to a more 'mainstream' operating model, 'in this sense, Iranian hip-hop cannot be analysed [...], because there are no alternative ways for hip-hoppers to survive financially' (2018: 262). The distribution system is certainly 'alternative', with some rappers burning tracks to CD-Rs and then, 'at the most populated junctions in the city, throwing a CD though the open windows of cars that had stopped at a red light' (Golpushnezhad 2018: 263).

The most striking aspect of this Iranian rap scene though is that the high level of state involvement in cultural life makes the distribution and recording of much rap music illegal, since it is seen as reflective of blasphemous, Western values. Recording studios have to be officially authorized; DIY rap practitioners therefore either record in hidden, (literally) underground studios, or they book studio time under the name of a state-approved pop singer who, should an investigator arrive, is on hand to quickly replace them in the vocal booth. In this way, 'state-supported institutions and state-sponsored cultural production facilities in Iran are gradually being co-opted and used for the creation of unofficial, DIY music and culture' (Golpushnezhad 2018: 269).

In some DIY scenes, then, there is a dimension of risk that is simply not present in UK DIY.[1] International research on DIY also shows that processes of globalization and localization play key roles in how, where and why scenes are formed. Quader and Redden's study of the DIY metal scene in Bangladesh shows that English-language schools operate as a key site for the consumption of international music, as well as the formation of musical communities (2015). Monika Schoop's recent work on DIY indie in the Philippines asks whether 'everyone is indie now', a concern that bears similarity with my own interest in 'DIY as default', but her answer to that question primarily addresses the persistent inequality and subsequent 'digital divide' in that country and in the wider world (2017).

Another distinction between different DIY scenes – and an important one, given this book's focus on social media – is that they can have very varied perspectives on the political and cultural value of technology. New tools for production and distribution can sometimes be perceived as having 'empowering' or 'activating' properties for DIY music scenes. Akane Kanai has framed 'remix culture' – a digitally enabled participatory form centred around the playful re-use and re-contextualization of existing cultural texts – as an example of this kind of DIY creativity (2015a). For me, a more convincing example of this technologically enabled cultural democratization is the UK grime scene which emerged during the early 2000s, and which made use of new digital music production tools as well as old distribution mediums – most notably community and 'pirate' radio in London (Hancox 2019). The history of this genre is littered with stories of teenage producers using 'cracked' (i.e. unlawfully copied) music software, in some cases to make hit tracks during breakfast before leaving for school (McQuaid 2015). But some DIY scenes find value in a scepticism towards new technologies, such as the Bay Area hip-hop scene studied by Anthony Kwame Harrison in the 2000s, in which cassette tapes were the preeminent medium for distributing music. The capacity for this outdated format to carry intertwined values of frugality and authenticity is encapsulated in a quote that provides his article's title: 'cheaper than a CD, plus we really mean it' (Harrison 2006).

---

[1] Or at least, not in the scene that is my object of study; police responses to Black youth music cultures in the UK could perhaps be seen as a comparable censorship (Fatsis 2019).

♪ 'Functions on the low' – **Ruff Sqwad**

So, DIY scenes across different genres and places can utilize very different practices. But there are also some key similarities. Autonomy, community and participation – some of the DIY values that I mentioned in the first chapter – are often invoked by practitioners across these scenes. There is also often evidence of a particular interest in labour and what kind of activity constitutes 'real' (or sometimes 'authentic') work. This sometimes relates to technology and its impact on 'traditional' practice, as evidenced in Ciarán Ryan's study of Irish music fanzine culture (2015) and also in Jhessica Reia's research on the Brazilian straight edge punk scene, where one practitioner derides 'younger people [who] have no idea of how to make a fanzine with photocopies and glue' (2014). Another continuity across most DIY scenes is a belief that what practitioners are doing is 'closer' to everyday life than commercial or popular music. 'Underground' means, for one practitioner in Kwame Harrison's US study, 'just regular people doing it' (2006: 290). DIY scenes also tend to emphasize their ability to generate values which are inviolably *internal* to their community, and which therefore cannot easily be 'converted' into external success or recognition.

Less positively, perhaps, there also seems to be a remarkable consistency in terms of the class composition of DIY scenes. Almost all of the researchers who have studied the diverse music cultures I have noted above – straight edge punk in Brazil, rap in Iran, metal in Bangladesh, the hip-hop tape scene in the United States, the UK 'anti-roads' and dance music culture – mention that the practitioners in question were mostly middle class or upper-middle class. Most of these researchers also offer some variant on the same explanation for this phenomenon: that participation in unpaid musical activity prerequires the uncommon capacity to give over one's time without financial recompense.

Three important historical DIY scenes

So far in this chapter I've suggested that while there are some similarities across DIY activities and lifestyles, there are limits to how much different versions of DIY music can be understood to have in common. Part of the reason for this is that, even in DIY, genre still matters. In several instances DIY can be seen as *extending* constructs of authenticity found in parallel popular music genres, their additional degree of organizational and structural control giving them the liberty to take steps to affirm authenticity which would be impossible from within the popular music industries. The three case studies mentioned below do this primarily by developing upon constructs from rock music, which I take to be part of a broader popular music culture. Emphasizing constructions of authenticity sheds new light on DIY's highly complex relationship with commodification, which I address further below.

In this section I offer a brief overview of three historical scenes – UK post-punk, US post-hardcore indie and riot grrrl – that might all be seen in some sense as forebearers to the 'indie-punk' scene that I observed in Leeds. Even at this level of specificity though, there is substantial diversity. These are scenes that interpreted the call to 'do-it-yourself' in different ways, whilst utilizing broadly the same base materials of guitars, records, tapes and zines.

These are not the only three historical DIY scenes that might serve to provide current practitioners with a sense of continuity. The 'twee pop' or 'C86' scene epitomized by the output of Sarah Records in the late 1980s and early 1990s was especially meaningful to some of my interviewees. Some people also emphasized the specific DIY history of Leeds, in terms of either its contribution to post-punk (and in particular to making post-punk 'fun' and 'danceable') or the city's engagements with other genres, including electronic music, noise and sound-art.

But, in considering the scenes below, I am not solely concerned with the direct impact and influence they have had on contemporary activity. The relative success of these three scenes means that they have been assessed and analysed by excellent writers (including some academics), and so they arrive helpfully pre-theorized, and this makes them particularly capable of acting as informal benchmarks for comparison at various points throughout the rest of the book. In making those comparisons, I am not suggesting that all or any of these scenes should be considered as the best or 'fullest' manifestations of DIY music culture, only that they serve to give a broader context to contemporary activity, in part by demonstrating the consistency of the tensions experiences and negotiated by DIY practitioners over several decades.

UK post-punk, 1978–83

Music journalist Andy Gill, writing in 1978, describes the first wave of punk rock as 'a kind of musical laxative'. 'Music cannot live on laxative alone', he continues, 'and the problem now seems to be one of what diet to pursue' (Gill 1978). In this context, post-punk music not only mounted an economic challenge to the major labels, it also questioned ideas of what pop music ought to be, what bands *were* and what they were for. DIY emerged in this period as one amongst many new models of music-making being trialled by practitioners hungry for new ideas.

♪ 'Messthetics' – **Scritti Politti**

Where punk had highlighted much of contemporary pop culture as boring and hypocritical, post-punk attempts to critique consumerism in this period are closer to Lukacs's conception of 'false consciousness' and Gramscian notions of hegemony, explicitly aligning themselves with a Marxist critique of the 'culture industry' and its role in maintaining societal passivity. This critique didn't always come from bands with a DIY approach; gestures of deconstruction and anti-consumerism are a stylistic feature of much post-punk in this period (e.g. XTC's self-satisfied, all-text album cover for *Go 2* (1978), which declares that album covers are 'TRICKS and this is the worst TRICK of all since it's describing the TRICK whilst trying to TRICK you'), and Gang of Four offers a particularly bleak vision of 'Entertainment!' released on EMI (1979). However, DIY bands were better able to connect aesthetics with action. DIY releases of this period (including by Desperate Bicycles and Scritti Politti) often came with pamphlets documenting itemized production and recording costs. For example, The Door and the Window's 1979 'Subculture' EP includes a flyer entitled 'How We Did It', showing costs including photo development and printing, recording, mastering and also including the areas where they avoided paying through their own activity ('collated sleeves ourselves') or favours ('recording equipment loaned by friend') (Ogg 2009: 131–2). The focus here is on transparency, particularly in an economic sense, as a means of breaking the commodity back into its component parts and *demystifying* a product that is generally presented as springing into being fully formed.

In doing so, DIY post-punk practitioners aimed to create a consumer product that might double as a self-help guide for the would-be producer. The Desperate Bicycles' second single contained an insert with the names of all the people who had contacted them about how to make a record, with the instruction 'now it's your turn' acting as a kind of 'calling out' of their audience to rise to the challenge and follow through on their initial enthusiasm (Selzer 2012). This positions consumers as producers by reversing the conventional temporality of cultural production, pre-emptively 'crediting' audience members for records they would hopefully go on to make. Attempts to offer transparency and demystification also took place at the organizational level. Rough Trade initially operated as a workers' cooperative in an 'unprecedented attempt to create internal record company democracy' (Hesmondhalgh 1997: 266); its founder Geoff Travis spoke of the label's desire to 'get rid of the idea that it's important to be a star, and to make the funnel wider, so as to include as many people and ideas as possible' (Birch 1979).

Post-punk DIY also attempted to negate apparent consumerist stupor through the discarding of some conventional elements of the pop song – the band Wire's manifesto includes rules such as 'no chorusing out' and 'when the words run out, it [i.e. the song] stops'. Gracyk notes that practitioners were working towards new styles of music, but moving in highly different directions, and 'until others imitated particular cases and, through copying, established a pattern of rules, no one could yet tell what those styles were' (2012: 83). This experimentalism links to an anti-populist, modernist tendency, but it also links to the participatory mode of production that is critical to DIY. If there is no 'right' way to play the music, then a lack of formal training need not constrain participation. Punk notoriously only required learning three chords before starting a band; post-punk experimentalism suggested you needn't learn any at all. But again, it is notable that the vast majority of this experimentation took place within the remit of the popular song – the desire to function as mass communication remains evident even as the components of the form were questioned and deconstructed. Post-punk though, for all of its aspirations to radicalize music, never successfully spoke to as broad an audience as punk.

US post-hardcore indie, 1983–8

Two excellent histories of US indie music, Michael Azerrad's *Our Band Could be Your Life* (2001) and Gina Arnold's *Route 666: The Road to Nirvana* (1993), identify a mid-period between 1970s punk and 1990s grunge (the two opposing ends of Arnold's titular 'road') as a golden age in which indie music flourished largely under the radar. Gina Arnold calls this scene 'Amerindie' (denoting a shift away from Anglocentric punk and new wave), and Azerrad simply calls it 'the underground'. I use the term 'post-hardcore indie' to highlight that its key practitioners 'passed through' hardcore punk, even as many ended up far beyond its restrictive genre boundaries. Hardcore was a very young scene – many of its participants being under eighteen – and as its practitioners grew up and grew apart, they took the 'do-it-yourself' ethos and applied it within new genres.

Black Flag, a California hardcore band formed in 1976, are credited as having 'built' the DIY touring network in the early eighties through their willingness to break new ground, taking chances in new towns and building relationships across the country. Labels like SST and Dischord, founded as local hardcore labels, became prominent indies within a scene that was stylistically much broader than punk; zines like *Flipside* and *Maximum Rock'n'Roll* (*MRR*) began with a focus on hardcore before likewise branching out. My analysis begins in 1983, the year in which hardcore punk, having emerged in Washington, DC and California circa 1980, appeared to many to be 'played out' (Andersen and Jenkins 2001: 166, Azerrad 2001).

Musicians and audiences began to question the more dogmatic elements of hardcore's style and sound, and there was a rapid acknowledgement and acceptance of other musical influences looking back beyond 1976 (punk's year zero) to country, psychedelia and classic rock.

In post-hardcore indie the live show generally had primacy over recorded output. Seminal act The Minutemen referred to all activity outside of the live show, including their recorded material, as 'flyers'; scene totem Ian MacKaye similarly designated his band Fugazi's records as the 'menu' and their live shows as the 'meal'. Physical recordings were often positioned as 'documents' of a band's current live sound, seeking to impart a cultural status more akin to historical archive (or perhaps Lomax's folksong collection) than ephemeral entertainment commodity. Ian MacKaye's first band Teen Idles had already broken up when in 1980 they recorded a single, meaning they had very little chance of recouping their costs through selling copies at shows, but their intent is summarized by MacKaye as: 'let's document ourselves' (Azerrad 2001: 132).

This documentational approach was extended to the recording studio. Bands recorded with very few overdubs and minimal studio effects, a 'clean' sound produced partly to reduce costs but also to ensure that records were an accurate representation of the performers' ability rather than an opportunity for technological experimentation. This positioned them in opposition to the apparent dishonesty of studio trickery which might raise a band above their 'natural' ability. It is notable that this scene co-existed alongside, and largely rejected, the growth in digital music technology in both professional and amateur contexts (Théberge 1997). In this way the distance between producer and consumer was purportedly minimized. DIY authenticity here is close to rock's emphasis on a 'no-nonsense' recording style, but the relative lack of commercial pressure within DIY allowed for this commitment to documentation to be taken further.

As the young hardcore movement matured and its key practitioners entered their twenties, many punks sought to distance themselves from the more destructive aspects of their scene. This meant not only a move away from physically violent behaviour, but also from the philosophy of refusal that characterizes what Moore (2004) calls 'deconstructive' punk. Several of the key figures in hardcore and post-hardcore came from military families, many participants were self-confessed 'nerds' who had a thorough understanding of electronics, and traits of rigor and attention-to-detail were highly valued. This is summarized by Faris, writing on Steve Albini, as a 'workingman persona' (2004: 432–4), which draws on American ideas of honesty and hard graft, reinforced by the everyman dress code of flannel shirt and jeans. Stability was highly valorized: Albini proudly identifies several indie labels as being among the most reliable and longstanding even in comparison to majors (Sinker 2001: 141). Mike Watt recalls making an extravagant display of setting up one's own gear, 'especially if you were playing with a mersh [commercial] band that had a crew and stuff' (Azerrad 2001: 74). The aim was to show that an ethos of personal responsibility and artistic integrity was not only an *alternative* to a system of contractual obligation and financial incentives, but that it might actually work *better*.

Both Arnold and Azerrad identify the importance of hippies as a lingering countercultural spectre haunting the US post-hardcore indie scene, and a desire to avoid that movement's co-option and reduction to stylistic touchstones as motivating a continued insistence on insularity. The often-forceful rhetoric against 'selling out' was grounded in a complex understanding of their position in history, and a conscious attempt to provide a new approach to an old problem. Arnold notes that 'for all that time, we were too ashamed of the fate of hippie idealism to recognize our actual allegiance to it' (Arnold 1993: 125).

Riot grrrl, 1990–96

Riot grrrl was a self-defined 'revolution' that began in Washington, DC in 1990. It was formed from within a punk scene in which women were often present, but frequently undervalued and disrespected – referred to as 'coathangers' by the men who would leave their jackets with them whilst they entered the pit, leaving the women 'literally marginalized' around the edge of the room (Koch 2006). Julia Downes describes riot grrrl as where 'young women attempted to disrupt the spatial and sonic norms of the indie gig to incite feminist community and provoke change in their subcultural situations' (2012: 205).

♫ 'I'm in the band' – **Bratmobile**

In their records, shows and especially through zines, riot grrrls attempted to open up new opportunities for women and girls to express themselves and to communicate with each other, calling for a 'revolution girl-style now' (Bikini Kill 1991). Riot grrrl became globally popular, particularly in the UK, with autonomous 'chapters' forming worldwide in order to coordinate local action. Their loud, fast punk music and confrontational performance style resulted in mainstream media coverage that emphasized their take-no-prisoners hostility, but alongside this anger was an emphasis on community-building ('girl-love') and tolerance towards difference. As well as being a specific way of doing music, it was also a specific way of theorizing and practising (third wave) feminism.

Riot grrrl was also incisively critical of mainstream culture's representation of women. Riot grrrls sought to distinguish their own output from other media commodities by positioning girl-to-girl communication as *vital* and *urgent*, an outcome achieved in part by using language that highlighted the dramatic dimension or the 'event-ness' of this process. Riot grrrl 'zines' and records are full of slogans, manifestos and calls-to-arms, creating an aesthetic of total urgency: 'BECAUSE every time we pick up a pen, or an instrument, or get anything done, we are creating the revolution. We ARE the revolution' (Reinstein, quoted in Dunn and Farnsworth 2012: 141).

The framing of riot grrrl as a means of giving voice meant there was a political value ascribed to action and production of any kind, encouraging others to be loud, to take up space and to communicate. The call-to-arms in the *Bikini Kill #2* zine suggests a near-uncontrollable refusal of hesitation: 'The undeniable genius of this generation has surfaced and it's all about ACTION, no time to decide what's right what's right what's right what's right' (Darms, Lisa and Fateman 2013: 123). Gottlieb and Wald find evidence of this within riot grrrl music, arguing for a reading of riot grrrls' screams as a rejection of the societal demand that 'women remain patient' (1994: 170). Riot grrrls consistently encouraged each other to produce, 'to take the initiative to create art and knowledge, to change their cultural and political landscape, rather than waiting for someone else to do it for them' (Garrison 2000: 154). The rhetoric of sharing also encouraged the extension of distribution networks through informal duplication using cassettes and photocopiers (Riot Girl #1, reproduced in Darms, Lisa and Fateman 2013: 31).

As Mimi Thi Nguyen notes, displays of emotional intimacy were key to riot grrrl's musical and social character: the highly personal nature of riot grrrl zines relates to the aim of feminist consciousness-raising and the idea that 'from inside the oppressed classes themselves come political knowledges based on experience, which might then be translated into expertise' (Nguyen 2012: 179). As much as mainstream media texts were an inspiration for the form of these zines, there was also an attempt to bypass their status as mass communication. The 'perzines' (personal zines) that were a key feature of riot grrrl make very few concessions to echoing traditional magazine content and style. Perzines are often closer to private forms of communication such as letters, or even diaries, attempting to create a mode of communication that is both one-to-one and one-to-many (Fife, 2019). Riot grrrl texts have an *epistolary* nature that renders them both mediated and unmediated – typified by zine-maker Nomy Lamm's assertion that 'I'm creating this kind of media that's literally from my most sacred place to somebody else's most sacred place' (quoted in Piepmeier 2009: 90).

One of Bikini Kill's mantras compels girls to 'struggle against the J-word [jealousy], killer of girl love', as part of a critique that identifies the individualistic pressures of the free market as well as patriarchal tactics that seek to set women into competition against one another (quoted in White 1992). Riot grrrl made considerable effort to deconstruct a pop hierarchy of 'star' artist and passive audience, and in the live setting bands would frequently offer the microphone to audience members in order to share information about upcoming shows and meetings, and to share experiences of sexism and abuse (Schilt 2003a). Key figure Kathleen Hanna claimed that 'with this whole Riot Grrrl thing, we are not trying to make money or get famous; we're trying to do something important, to network with grrrls all over, to make changes in our lives and the lives of other grrrls' (Hanna, quoted in Dunn and Farnsworth 2012: 11).

DIY and popular music: An ambivalent relationship

In attempting to define DIY musical culture, scholars have often emphasized its *difference* to, rather than its similarities with, popular music culture in general. This difference is frequently identified in terms of 'resistance' (Schilt 2003b, Duncombe 2008, Downes 2012, Guerra 2018), where the social and economic organization of the scenes in question is understood to constitute an 'other' to a proposed hegemonic structure of cultural power. DIY is also presented as a kind of social movement; in these accounts it is defined by its close connections to extra-musical attempts to re-shape society or to attempts to live in a 'counter-cultural' or 'oppositional' fashion (Dunn 2016, Radway 2016, Culton and Holtzman 2010). Relatedly, there is a notable tendency to theorize DIY music in terms of its success (or otherwise) as a form of radical political praxis. Paul Rosen considers DIY as 'an example of anarchism in practice' (1997), and Pete Dale tracks the consequences of competing Marxist and anarchist tendencies in DIY (2012).

In my view, an over-emphasis on conceptualizing DIY as social movement, or as political praxis, significantly neglects one of its defining features: namely, its close *emulation* of popular music culture and the organizational forms of the commercial music industries. Even if we accept that DIY is perhaps 'resistant', the specific *character* of its resistance cannot be understood without accounting for this ambivalent relationship to popular music, which is consistent throughout DIY's various manifestations – at least in the 'indie-punk' varieties I have just outlined.

DIY music has a close coherence with, and affinity to, popular music forms, texts and infrastructures, and this is a critical part of its character. The key cultural units of pop and rock music – the live show, the record, the band, the label, the audience (as well as more modern additions: the music video and the playlist) – are similarly the key units of DIY. Whilst there have been attempts to deconstruct or subvert these concepts, they follow mainstream pop music inasmuch as they work not only *through* mass media, but *as* mass media. DIY music is a response to the pitfalls of commodification and media power which deals primarily in *commodified, mediated communication*. This doesn't make DIY politically redundant, but it does imbue it with a particular 'shape' and character.

This point can be emphasized by noting DIY's substantial *differences* to other amateur musics which seem more clearly to have a participatory character. One particularly politicized example to this might be found in UK 'street choirs' – a long and important history recently captured by the Campaign Choirs Writing Collective (2018) – but more generally in a wide array of participatory musics in which distinctions between performer and audience are dissolved or non-existent (Turino 2008). Given that DIY is purportedly deeply interested in increasing participation and minimizing artist–audience distinctions (Verbuč 2018), it is notable that it very rarely takes an approach which thoroughly emphasizes participation over and above adherence to the forms and units of popular music.

DIY's ambivalence towards popular music must be seen as resulting from an acknowledgement and appreciation of the communicative power of popular music, and its particular political potency. Notions of authenticity, rebellion, social upheaval and speaking truth to power have been encoded in popular music from at least the 1950s (Keightley 2001, Frith 1996), and when DIY identifies popular music as an instrument of social change, it is drawing upon lineages that are very much 'within' the mainstream music industries, as well as upon more radical political and cultural lineages.

Marxist readings of DIY tend to hinge on its capacity to in some way 'de-commodify' music. However, framing the issue in terms of a 'punk/ commodity opposition' (Thompson 2004: 81) is unhelpful; accounts which emphasize DIY's capacity to resist commodification often rely on a kind of special pleading or a rather shallow definition of commodification.² As I have shown above, the commodity form of recorded music has proven itself to carry huge cultural and political potential, and that aspect of its exchangeability clearly holds an appeal for DIY practitioners which they are reluctant to lose. If practitioners were concerned about commodification *above all else*, then, as I have mentioned, there are participatory forms of music on offer that would seem to be far less threatened by commodification.

DIY is therefore best understood not as a form which attempts to radically overhaul the organizational and cultural units of popular music,

---

²Thompson, for example, argues that Crass' musical output is anti-commodification because it avoids radio-friendly song structures, but the same is not said of the avant-garde music (e.g. progressive rock, early electronic music) being made at the same time in other realms (2004: 84); record collections are fetishistic except when owned by a punk modelled on Benjamin's 'true collector', who can re-individualize through their ability to recount a 'life history' (2004: 124).

but as one which attempts to 'fix' perceived problems with popular music's role and position in society. The aim is to shift the terrain in some way, without seeking to argue with the fact of pop music's communicative power: 'pop music … *but better*' – where 'better' might stand in for any number of specific adjustments required to create a popular music which is in keeping with the aims of a given scene.

In the next section, I suggest that DIY's ambivalent relationship to popular music doesn't necessarily discount its capacity to offer something akin to 'cultural resistance'. But this ambivalence does result in some very specific and irresolvable tensions, which are fundamental to, and indeed *constitutive of*, DIY music. Three of these tensions are outlined briefly here, and their ramifications recur throughout the rest of the book.

The first tension relates to production and consumption. DIY holds, generally, that cultural production is a form of power, and that the existing structures of cultural production both represent and constitute an unequal and problematic power balance.[3] It therefore aims to encourage wider participation in the production of musical culture: anyone can do it, and everyone should. However, since DIY also carries a strong belief in the power of the recorded music commodity – the seven-inch single as a life-changing phenomenon – the role of the *consumer* remains prominent in DIY in a way that is not the case for many other participatory musics. So, DIY is faced with the question: what is so very special about the producer–consumer relationship in this instance? In what ways are consumers of DIY music understood as similar or different to conventional music consumers? DIY practitioners respond to this tension by creating commodities that attempt (successfully or otherwise) to bypass or mitigate consumption's connotations of passivity, exploitation and alienation. This might be attempted through a myriad of approaches including aesthetics, performance modes, organizational structures, or production and circulation strategies.

A second tension relates to how open or closed a given DIY scene ought to be – that is, the extent to which it ought to use its mediated communicative capacities to try and make an impact beyond its borders. Some DIY scenes have emphasized the need for a safe haven in which political and aesthetic ideas might develop; others have found value in opening up the scene to confrontation and debate. A focus on insularity sometimes leads to accusations of elitism and irrelevance, but when DIY scenes are seen as insufficiently distant from mainstream culture, they sometimes struggle to offer a convincing challenge to existing cultural norms. This need not necessarily be seen as a tension between internal values and external rewards

---

[3]In this regard DIY shows an affinity with the concerns raised by prominent mid-twentieth-century critics of the 'culture industry' (Horkheimer and Adorno 2002, Marcuse 1991, Packard 1957), as well as with macro-historical understandings of consumption as passive and/or wasteful (Miller 2001: 2–6).

(i.e. the 'selling out' paradigm), but as differing perspectives on how far the mediated commodities of DIY can travel before their political value begins to diminish.

If that tension between insularity and openness might be considered as a 'spatial' issue – that is, how far to travel – then it is also necessary to note a comparable 'temporal' tension, relating to the speed at which DIY activity unfolds through time. Popular music is in many senses a routinized industry, and DIY practitioners find value in finding an alternative to that routine, although they may disagree on whether to work faster or slower. Anthony Kwame Harrison's DIY hip-hop practitioners were immensely prolific – one artist had recorded seven albums in two years – and saw that as an anti-commercial practice of 'making and releasing music whenever you feel inspired', rather than sticking to more business-savvy practices of 'releasing one album every two years' (2006: 292). In Ciarán Ryan's study of Irish fanzines, it was the unhurried nature of production that practitioners valued, in comparison with blogs (then a relatively new self-publishing option), which they saw as locking authors in a detrimental schedule of frequent publishing (2015: 251). This temporal tension often relates to a larger existential concern. Active, busy DIY scenes are often ones that are optimistic regarding their potential to make an impact on the world. Some scenes, though, might question the very nature of action and productivity, seeing all activity as grist for the capitalist's mill, and cultural production as *a priori* commodified and exploited – in these instances practitioners might locate resistant value in a slower or more cautious approach.

DIY music as cultural resistance

DIY music is often framed, by scholars and practitioners alike, as a kind of cultural resistance. It is understood as standing, consciously or not, in opposition to some dominant (or 'hegemonic') values, and in opposition to some of the institutions responsible for perpetuating those values. In the broader field of media scholarship though, the notion of resistance is generally seen as a problematic simplification. Media scholars Abercrombie and Longhurst refer to this as the 'Incorporation/Resistance Paradigm', which they contextualize as a specific historical moment in media and cultural studies. In this paradigm, texts are either taken 'as read' (in which case hegemonic power goes unchallenged) or they are 're-coded' by audiences in ways that might resist hegemonic power. Abercrombie and Longhurst argue that this paradigm no longer reflects – if it ever did – the extent to which media has 'leaked' out into everyday life and the way in which media texts have become 'intimately bound up with the construction of the person' (Abercrombie and Longhurst 1998: 37).

It is true that 'resistance' does not fully capture the variety of ways in which cultural activity can be meaningful and valuable. Participation in 'the arts' in general can have a wide range of benefits in terms of both personal well-being and social cohesion (Matarasso 1997); Thomas Turino similarly espouses the individual and collective benefits of musical participation specifically (2008). Hesmondhalgh and Baker's model of 'good work' in cultural labour emphasizes its capacity to offer 'autonomy, interest and involvement, sociality, self-esteem, self-realization, work-life balance and security', as well as the importance of making 'good' products that might also 'promote aspects of the common good' (2011: 36). This demonstrates that a methodological focus on the 'good' need not preclude engagement with weighty political and social questions, and these accounts draw attention to the positive aspects of cultural activity that a focus on resistance might tend to understate. This focus on the 'good' is evident in some recent work on DIY, too: Evangelos Chrysagis argues that the Glasgow DIY scene is not 'predicated upon what is usually called "resistance"', but upon positive practical action (2016: 293). DIY, he says, feels for most people like productive 'doing' rather than 'negating' some vaguely defined antagonist.

So, what does 'resistance' offer analytically that a focus on the 'good' might not? The first benefit is a sense of some cultural activity as being, in Stephen Duncombe's terms, a 'stand against' (2017: 176). This is also important in forming the qualitative character of the relationship between DIY practitioners and various other groups (addressed in Chapter 5): a sense of oppositionality is in part what differentiates DIY from the related but distinct field of community arts. The second, related, benefit of utilizing 'resistance' is that it suggests that this oppositional position might serve as the basis for specific forms of collectivism. Hesmondhalgh and Baker's focus on 'the common good' points towards this dimension, but doesn't address the sense that other people might also be working towards those ends (and that some social groups might not be). DIY music does not constitute a social movement – its aims and practices are too varied, too contradictory – but a focus on resistance points towards DIY's movement-esque qualities, particularly in the way that DIY practitioners might feel connections of solidarity between themselves and other practitioners. I take resistance on board, then, to emphasize DIY music's specific connections to alterity and collectivism, and to affirm that these are worth retaining – whilst acknowledging that resistance is not the only valuable outcome of DIY activity.

Stephen Duncombe's work on DIY culture offers a fairly comprehensive framework for assessing the kind and scale of cultural resistance, influenced by his experiences as a punk practitioner and political activist. Duncombe's framework is based on assessing several 'scales of resistance': a scale of political engagement (from unconscious to self-conscious), a scale measuring the social unit involved (an individual, a subculture or a society) and a scale measuring results (survival, rebellion, resistance). These sliding scales are useful in considering how forms of resistance might depend on the social unit in question (i.e. individual or group). At the individual level, resistance is linked to notions of self-realization, empowerment and autonomy; towards the other end of the scale (bigger social units, a 'larger' kind of resistance) is the building of alternative economies. At its most aspirational this approach calls for culture to be repurposed and reorganized in order to bring about a political revolution.

I am a little more reticent than Duncombe to identify DIY and 'subcultural' activity as the sole or primary avenue of cultural resistance. Duncombe argues that 'in a society built around the principle that we should consume what others have produced for us, throwing an illegal warehouse rave or creating an underground music level – that is, creating your own culture – takes on a rebellious resonance' (Duncombe 2002: 7). However, these 'underground' environments often fail to acknowledge injustices and uneven power dynamics within their borders; Julia Downes, in her ethnography of UK riot grrrl, notes that 'girls and young women frequently fall short of achieving the authenticity and legitimacy dictated as necessary for full participation within subcultural spheres' (2009: 26). An emphasis on the 'underground' also underplays the complexity of media landscapes that even the most subcultural practitioner moves through, and the new ambiguities regarding distinctions between production and consumption that are a key focus of my study.

These complexities are underserved in much literature on DIY as (sub) cultural resistance. There is often an unspoken conflation of 'underground' culture with progressive politics, which is further complicated by a tendency to read leftist politics as more authentic or more deeply felt than other political positions: when grassroots culture is leftist, it represents the surfacing of the common political will, and when it is nationalist or populist, it represents the uncritical regurgitation of conservative mass-media discourse. A clearer articulation of the politics of resistance would allow a more thorough assessment of problematic or unhelpful elements within activity inscribed as resistant, such as the individualist libertarian implications of Thoureau's 'civil disobedience' (1986), or the racism and sexism observed by John Clarke in his study of British working-class 'skins' culture (2006). Closer to home is the racism and misogyny that has sometimes – *often* – been a part of DIY music scenes and which continues to be a serious problem today.

Clearly then, if the notion of cultural resistance is invoked as part of any *normative* claim – that is, an argument about the way things ought to be – it is not sufficient to take the scene's own word for it and to merely trust that DIY resistance will organically align with any particular political project. Indeed, DIY's particularly self-conscious sense of itself as 'resistant' generates its own effects, which may not always be beneficial. The way it is sometimes described in terms such as 'music for misfits'[4] has, I think, permitted a self-construction as 'outsiders' which can be elitist and vengeful.[5] DIY is a marginal music, but it also *produces* its own marginality. It then valorizes this marginality which, in my view, is not in itself a politically valuable aspect. (Lots of political and cultural groups are marginal, many of them holding very different principles and practices to those found in DIY.)

So, we need to be willing to submit DIY activity to examination and interrogation, and to provide a sturdier means of assessing any relationship between resistance and injustice, and including the positive or negative dimensions of that relationship. To this end, I draw upon the socialist-feminist political philosophy of Nancy Fraser. In her work on social injustice, she identifies of 'misrecognition' and 'maldistribution' as key categories of injustice that constitute twin blights upon Western society under late capitalism (Fraser 2000). Fraser demonstrates that these issues of redistribution and recognition are not reducible to the simplistic 'folk paradigms' of 'class politics' and 'identity politics', which in these forms often appear to be mutually exclusive (2003: 11). Combining Marxist and Weberian conceptualizations of societal divisions, she argues for an understanding of 'two-dimensional subordination', whereby subordinated groups 'suffer both maldistribution and misrecognition *in forms where neither of these injustices is an indirect effect of the other, but where both are primary and co-original*' (Fraser 2003: 19). Importantly, Fraser identifies that recognition is not merely an issue of interpersonal ethics, but also relates to being recognized by 'social institutions'. Maldistribution and misrecognition both occur 'when institutions structure interaction according to cultural norms that impede parity of participation' (2003: 29).

Fraser's model of social justice suggests that power might operate along both economic and cultural lines, and provides some insight into how these two lines intersect as in, for example, the ways in which the cultural industries maintain economic inequality *and also* provide limited opportunities for representation. It also highlights the critical importance of addressing these two dimensions simultaneously. This is not to say that they cannot be beneficially addressed individually, but rather that addressing

---

[4]*Music for Misfits* was the title of a recent three-part BBC television documentary on UK indie music – the first episode of which focused on 'The DIY Revolution'. There have been countless presentations of indie and DIY that place a similar emphasis on marginality.
[5]Sometimes this leads practitioners to over-identify as victims of hardship and oppression in ways that, when not laughable, can be offensively wide of the mark. One prominent example of this is the Ian Mackaye-penned lyric to the Minor Threat song 'Guilty of being white' (see Duncombe 2011: 99–105).

one of these goals gives no guarantee of positively progressing the other, and may even work against it (Fraser 2009). And, as Fraser notes, drawing on intersectional feminist theory (Crenshaw 1991), 'individuals who are subordinated along one axis of social division may well be dominant along another' (2003: 26).

This helps us to see that DIY ethics can be read as similarly two-pronged – with different historical scenes placing different emphases on the importance of redistribution (i.e. political economy) and recognition (i.e. politicized identity). But more importantly, Fraser gives us a normative framework – albeit a loose one – by which to relate DIY cultural resistance to some external principles of social justice. It is intentionally broad-brush, at least in my application, and leaves space for considering the multiple approaches by which social justice aims might be met, as well as the times and places where resistance may not correlate to these aims at all.

There are a couple of aspects of Fraser's framework that might benefit from some attention and clarification, both of which relate to subjectivity – that is, the feelings attached to being one's self, and the political dimensions of this experience of 'self-ness'. Fraser's 'misrecognition' does include the potential for misrecognition of the self, understood as the ways in which institutions might impede a given subject's capacity for self-recognition. But this is an awkward phrasing, which seems to rely on the liberal notion of an essential, true self that can be 'recognized' without reference to the social world. As such it is somewhat unsatisfactory for dealing with the complexities of 'self-governance'. This is the notion that the power we exercise over ourselves is structured and shaped by other powers which, even though we may not realize it, are leading us to seek certain ends rather than others. Thus, even when we aren't being directly told what to do, we 'govern' ourselves into being certain kinds of subjects.

The French philosopher Gilles Deleuze considers self-governance in terms of a shift from 'societies of discipline' to 'societies of control', which he argues has taken place since the end of the Second World War (1992). In the former, it is state-affiliated institutions (schools, hospitals, prisons, religious institutions) that do the work of 'molding' subjects; in the latter, subjects are 'modulated' by post-institutional controls that reward those who have internalized appropriate forms of motivation. The individualized subject of late modernity is, as Wendy Brown summarizes, 'quintessentially susceptible to disciplinary power', since 'their individuation and false autonomy is also their vulnerability' (1995: 19). Through this critical lens, the very terms on which DIY seeks to ground its own capacities – terms such as autonomy and empowerment – are called into question. This relates closely to the concerns outlined in Chapter 1, regarding the changing meaning of self-sufficient cultural activity. Importantly, it suggests that the influence of powerful external actors (such as social media platforms) may not always feel as though it is arriving from 'outside' of ourselves. But at the same time, we need not see a certain permeability between self and society as inherently negative (else we fall into the liberal error of assuming the self to be, ideally, a standalone entity). Rather, such permeation is surely inevitable and works both ways. So, whilst it augments the relationship between cultural resistance and social justice, the notion of self-governance does not entirely tyrannize the ethical dimensions of DIY activity.

3

The personal

Subjectivity and self-representation on social media

For the practitioners I interviewed in Leeds, a huge part of DIY music's appeal was its capacity to offer experiences of self-realization. Making culture (particularly music, but also zines, visual art and other cultural texts) was often seen as 'transformative' in terms of developing DIY practitioners' confidence and capabilities, and as contributing to a rich and rewarding inner life. This feeling of self-realization was often linked to 'firsts' – many practitioners had strong memories of forming their first band, playing or promoting their first show, or of writing and recording their first songs. One illustrative example of this comes courtesy of a female practitioner who had been involved in the Leeds indie-punk DIY scene for a number of years in various roles, but who had only just started performing music herself. She spoke powerfully about the (recent) experience of playing her first ever show, in her first ever band, playing an instrument she had only just begun learning. She recounted this experience in the form of an affective journey, at both an individual and a collective (i.e. band) level:

> We've psyched each other up, basically, to do something, and that's how [the band] formed. We're all just really anxious people so it's quite … not nice, but comforting to know that we're all in the same position, none of us are uber confident. […] The first show was terrifying. I thought I was gonna be sick the whole time. In the day [before the show] I turned into a different person, I was really snappy and weird and I didn't know why I felt so strange, but it turned out I was just really anxious, cos once we

played I felt this weird relief that I'd never felt before and I was like 'that's nerves'. I didn't know I could ever feel that nervous about something, basically. It was terrifying but it was good cos we played in a little tiny room and there were quite a lot of our friends there, maybe twenty, so it was fine, everyone was really supportive, and I was like 'ok – maybe I can do this'. (P19)

I think most practitioners would recognize and identify with some aspect of this recollection, which reflects something like DIY 'at its best'. DIY here provides an experience of crossing a threshold, and thus of recognizing one's capacity to achieve something that may once have seemed impossibly distant. This belief that DIY music activity can enable self-realization doesn't necessarily have to involve a comparison to, or critique of, the music industries – although it sometimes does. It is an inherent acknowledgement that musical work has some particularly valuable traits. Yet, in offering those traits to a wider range of people and, in effect, making musicians out of non-musicians, DIY can call the nature of those social categorizations into question.

Practitioners also saw DIY as allowing a freedom to truly 'be one's self' through creative output, in a way that they thought might not be feasible for musicians working with a greater commercial imperative. Accordingly, the apparent 'quality' of the end product was often considered less important than the processes and social relations that operated around its creation. One Leeds DIY practitioner reflected:

I think the whole good/bad musician thing has been something which used to hold me back, but now it doesn't [...] now I am concerned with just making things, and the process of making stuff yourself as this really important, transformative process. (P21)

The term most often used to describe this kind of process was 'empowerment'. This is something of a fraught term, which some critics have suggested validates individualistic self-improvement at the expense of activity that wields power in the name of wider change. For example, the political philosopher Wendy Brown writes that

contemporary discourses of empowerment too often signal an oddly adaptive and harmonious relationship with domination insofar as they locate an individual's sense of worth and capacity in the register of individual feelings, a register implicitly located on something of an otherworldly plane vis-a-vis social and political power. (Brown 1995: 22)

Empowerment is certainly a slippery concept, capable of being put to various ends. However, DIY's emphasis on music *as communication* – partly a consequence of its ambivalent emulation of mainstream popular music, outlined in Chapter 2 – means that empowerment here is not as solipsistic or as 'otherworldly' as might be assumed. It is empowerment that must be seen and heard; only in this way can it function as an encouragement to others, in accordance with the 'anyone can do it' ethic of DIY. As such, it becomes entangled with concerns over how to best communicate, using music and media, the truthfulness of the empowerment being evidenced. In this effort to create and codify an aesthetics of 'really meaning it', DIY shows itself to be – like so many other kinds of popular music – deeply concerned with authenticity.

The first section of this chapter assesses the importance of authenticity – and particularly an 'intimate' authenticity – in DIY music cultures. Drawing on interviews with DIY practitioners in Leeds, it considers the impact that social media's recalibration of communication has had on the presentation of 'authentic' DIY subjectivities. I suggest that the 'epistolary intimacy' that was previously a distinctive communicative style of DIY is now increasingly normalized on, and perhaps even compelled by, social media platforms. The second section develops these ideas further, by examining how social media's impetus towards relatability intersects with contemporary politics of visibility in DIY.

The third section looks at the 'everyday' aspects of being on social media – that is, the large amount of time spent 'on' platforms that doesn't seem to specifically relate to DIY practice. I argue that such activity, and practitioners' feelings about it, are highly relevant to gauging social media's impact on DIY ethics. The extent to which practitioners feel commodified and surveilled as social media audiences necessarily impacts their capacity to feel 'resistant' when producing on and distributing through those platforms. I find there is little room for positivity in this regard; practitioners are pessimistic about their role as platform consumers and struggle to find appropriate modes of non-participation that would not also cut them off from the rewarding elements of platforms.

Intimacy and authenticity

In the previous chapter I suggested that DIY music has an ambivalent relationship to commodification, and that 'authenticity' – a well-established concept in popular music studies[1] – is the most appropriate means by

---

[1] Popular music scholar Simon Frith has called authenticity 'the most misleading term in cultural theory'. It's misleading, he argues, because our focus should not be on measuring music's proximity to 'truth', but rather on understanding 'how it sets up the idea of "truth" in the first place' (Frith 2007: 261). This is the sense in which I use the term – as a means of considering the strategies by which music makes and communicates truth claims.

which to consider how this ambivalence manifests. In DIY, the evidence of commodification is generally not identified through economic analyses, but through subjective aesthetic judgements – whether or not the music *sounds* standardized and excessively mediated, or whether it sounds truthful. Issues of commodification intersect closely with issues of taste, in the labelling of music as good or bad, as 'real' or 'fake', as interesting or boring. Thus, in seeking to minimize the pitfalls of commodification, DIY practitioners have generally sought to produce particular *types* of commodities, which can still act like commodities in terms of use and exchangeability.

There are many different, overlapping models of DIY authenticity in operation, and individual practitioners often have very different perspectives on genre, aesthetics and politics. In this section, though, I focus on the presence of a relatively coherent 'intimate authenticity'. DIY does not have a monopoly over intimacy, of course. Singer-songwriters (including stadium-filling performers like Ed Sheeran) claim a comparable kind of intimate authenticity premised on a stripped-back, 'no frills' aesthetic intended to suggest that a one-to-one communication can take place even across vast, mediated distance (Till 2016). But DIY's notion of intimacy often relates to specific concerns regarding the consequences of commodification.

Commodification means prioritizing exchangeability over specificity,[2] and this is sometimes seen as detrimental to art's capacity for self-expression. The music industries' emphasis on market exchangeability (i.e. capacity to make profit) leads to an emphasis on making uniform 'standard' products (thereby lowering production costs), and this plays a role in generating aesthetic conservatism too (since the more novel the product, the greater the risk that consumers won't take to it).[3] One key concern, then, is *standardization*; accordingly, DIY commodities tend to be presented as non-standard. The classic example of this is the handmade or hand-replicated music commodity, whereby the small textual and material differences that emerge from this domestication of industrial processes are as often presented as evidence of authenticity.

---

[2]In the first chapter of *Capital*, Marx lays out a straightforward definition of the commodity: an object is considered a commodity when its use-value (the purpose it is specifically for) is subjugated by its exchange-value. However, the process of commodification is rather more dispersed. Marx understands this as a form of alienation resulting not from the single commodity's entry into the market, but from the general tendency to consider objects (and also, eventually, people) primarily in term of what they can be exchanged for, that is, their economic value (Marx 1976: 48–9).

[3]Since music commodities effectively *require* some (minimal) novel element, the music industries have often granted musicians relatively high levels of autonomy over their labour (Stahl 2013: 1–2). Labels' managerial workers act as a 'buffer' between the 'capricious creative' and 'corporate accumulation imperatives' (Banks 2007: 9) to channel creativity into profitable products.

Another concern is that commodification tends to result in products being perceived as something other than the product of human labour – Marx's notion of 'commodity fetishism' (1976: 165) – and thus come to seem preordained or mystified, as though they had arrived from another world. DIY practitioners have historically tended to be ambivalently interested in this mysterious quality, but also in the notion that we can, effectively, 'make our own mystery', and that this capacity is denied when cultural texts are excessively mediated (i.e. made more distant than they should be from our everyday life). A handwritten or handmade aesthetic has often been perceived as countering this mystification, too (especially in zines), and some historical DIY scenes have also emphasized the importance of a 'lo-fi' (or 'bedroom') recording aesthetic in this context. These techniques both seek to create cultural texts that, in a sense, bear traces of their own production process. This aesthetic strategy of rendering the *act* of zine-making visible in the zine, or the recording process audible within the song, historically constituted an effort towards demystification intended to demonstrate that cultural production was achievable for anyone. Many contemporary DIY practitioners still make zines, and one participant described their zine-making (and, less often, their music-making) in such terms:

> I want it to kind of look like anyone could do it, to encourage other people to do it. And similarly, when I play in bands I'm not a very competent musician, but I still think the music I make is worthwhile, and that people will enjoy seeing it. And I guess that's part of DIY for me, that I don't feel like I have to reach a certain level of proficiency before I'm allowed to share my art with other people. I can just do it when I want to. (P9)

This is in keeping with DIY's broad interest in demystifying cultural production, but also resonates with a longer historical interest in the democratization of culture. For example, speaking in 1934, the cultural theorist Walter Benjamin highlighted the Soviet Russian socialist press as the kind of cultural output that 'forces us to re-examine the separation between author and reader', thereby creating a society in which the reader is 'always ready to become a writer' (1970). It is this kind of propulsive political potential that practitioners sometimes locate in intimate, lo-fi or handwritten cultural texts.

Continuing to utilize this kind of DIY aesthetic can still constitute an attempt to demystify the production process, but arguably retaining a 'handmade' aesthetic means adherence to an established 'style' of DIY culture that no longer relates to the recent technological developments that have made cultural production more accessible. Physical cutting and pasting, photocopying and so on, whilst convenient and cheap for punks and riot grrrls of the past, are no longer strategies that adhere to the 'it was easy, it was cheap' philosophy of DIY. Social media platforms offer an easy way – *the* easy way – to distribute and present DIY activity.

However, the capacity for online communication to bear traces of its own production process (as zines and lo-fi recordings arguably do) seems limited. For example, Facebook Pages – a key tool for almost all DIY bands in Leeds during my fieldwork – equips practitioners with web presences that are near enough identical to one another, with the exception of cover photos and profile photos. Consequently, these platform templates tend to look clean and professional. They are a far cry from the consciously non-professional aesthetics of earlier iterations of indie-punk DIY. They are also vastly different from previous iterations of personal webpages: the aesthetic of 1990s home pages and even the proto-platform aesthetic of early 2000s MySpace were both characterized (for better or worse) by highly variable designs, borne of unmistakably autodidactic coding (see Arola 2010). One Leeds practitioner reflected, with some ambivalence, on the consequences of Facebook Pages' professionalizing aesthetic as it related to her DIY music skill-sharing group:

> You know … [compared to] riot grrrl, it's actually got more reach [on social media], we reach more age groups than we would the old school way, [there's a specific] kind of person who would pick up a hand drawn zine, whereas someone is more likely to come across a Facebook Page and message. Because this idea of what professionalism is supposed to look like goes away on Facebook, almost? Cos it's just text on the screen, not handwritten text. Which in my handwriting would not be so good. So, we get like producer-y, high level people, and then we get the nine-year-old who's like, 'my dad and I came across your page', that kind of thing. It is kind of weird, cos it loses personality right, we all are the same, whereas I could change the way it looks on the [written] page, but [online] it's like we all have a profile pictures and a cover photo and we all have these timelines. Because we're all the same. Like, if we all are nothing except for like … we're 'fill-in-the-blank' people, your name and your photo, then it's kind of egalitarian, no-one can actually be bigger than anyone else, no one can have a Facebook Page that has more data than the other, we're all allotted this same space, and we all can use it just as much as the other. (P28)

This suggests that the imposed design of Facebook Pages (and other social media 'templates') might offer both democratization *and* homogenization, by removing the distinctive – yet potentially exclusionary – subcultural capital imbued by a handmade aesthetic. DIY culture in this way is made accessible to a wider set of people, allowing its values and politics to travel further. But at the same time, a DIY aesthetic is overridden by a platform aesthetic.

Media scholars Alison Hearn and Sarah Banet-Weiser have suggested that this platform aesthetic carries an element of 'glamour' that means, in a sense, the platform is the star (2020). Drawing also on Ed Finn's work on platforms' 'aesthetic of abstraction' (2017) – that is, the tidying away of complexity behind glossy, seamless interfaces – Hearn and Banet-Weiser suggest that this glamour assists processes of 'beguilement and obfuscation' which mask platforms' economic power (2020: 9). So, whilst the benefits of wider communicative reach identified by the DIY practitioner above are undoubtedly real, the obfuscating role of platform aesthetics might serve the precisely inverse purpose of DIY's use of intimate aesthetics to *highlight* and encourage agency.

Another key facet of intimate authenticity is that it can signify that the communication between producer and consumer is of a particular, non-mediated kind (even whilst using media to communicate). This was particularly valued by riot grrrls. The zines they created usually had no editors and, unlike the glossy girls' magazines of the time, had no corporate owners or advertising partners to answer to, and therefore could speak honestly and openly about political and personal issues that would otherwise be under- or misrepresented. As outlined in Chapter 2, riot grrrls used an 'epistolary' (i.e. akin to letter-writing) style of communication to create a sense of emotional intimacy as well as a politicized honesty between themselves and their audience (Nguyen 2012).

The internet is very often seen as a place where these kinds of intimate cultural communications can occur. What could be more epistolary (and yet still a form of 'mass communication') than a blog, or a vlog or the sporadic missives that characterize Twitter output? The disintermediation enabled by platforms would seem to permit communication that travels – to reiterate riot grrrl zine-maker Nomi Lamm's claim from the previous chapter – 'from my most sacred place to somebody else's most sacred place' (Piepmeier 2009: 90).

But if an aim of DIY has been to make public communication more personal, social media has made personal communication increasingly public. Millions of us communicate 'intimately' in various public online fora every day, and the very term 'platform' suggests the public elevation of this kind of communication (Gillespie 2010). So, in keeping with the broader shift identified in this book, there has been a move from 'anyone can do it' to 'everyone is doing it'; intimate self-expression through media is no longer the sole preserve of DIY culture. This shift – as with any broader adoption of DIY cultural practices – does not necessarily rob intimacy of its valuable qualities, nor does it inherently mean that the self-expressive potential of DIY has suffered. It does suggest, though, that we should pay close attention to precisely how and why an intimate mode of authenticity might be articulated to new or developing social and economic processes.

One question might be whether DIY practitioners, by virtue of their independence, are granted any *more* capacity for this kind of 'authentic' communication – that is, whether they can be more intimate, or more honest, than a musician with commitments to a major record label, and particularly with regards to issues relating to social justice. One Leeds practitioner did argue that their independent status meant they had more freedom to speak out on social media, therefore maintaining a distinctly DIY authenticity: 'I like the fact that we can just freely retweet a political post without a label coming and saying, "you can't do that, it will affect your sales"' (P16). However, this kind of intervention or censorship would be fairly unlikely today – record companies *want* intimate, self-expressive social media content from their rosters and are happy to permit controversial content (up to a point) in exchange for social media's capacity to generate attention and press coverage. In fact, DIY practitioners are, if anything, now playing catch-up with celebrities, and often with the rest of the public, when it comes to displaying this kind of intimacy. Intimate authenticity has become a prevalent feature of popular culture, a trend that has been echoed in the growth of reality television, reflecting a strong desire to get in 'behind' media artifice in order to know what people are *really* feeling. (Or at least, what it looks like they are really feeling.) And just as reality television emerged as a low-cost model of cultural production (Hearn 2017), expressions of authenticity come at a bargain price – that is, free.

♪ 'Transparency is the new mystery' – Marnie Stern

Another important question is whether platforms' communicative norms might impose limits on the kinds of intimate authenticity that are communicated or which tend to be seen as valid. Several of the Leeds practitioners I interviewed were convinced that they didn't have the right kind of personality to enjoy Twitter, or to be successful on it. They didn't feel comfortable with that kind of expression or, often, weren't sure how to translate their feelings into enjoyable or relatable content. One said they didn't use Twitter because 'I don't have that ironic persona that far-left Twitter has, that everyone seems to go crazy for' (P17); another assured me that they were 'absolutely crap at Twitter – don't ever look at my Twitter' (P19). To be successful on social media requires aspects of the self to be made readily available (and, usually, made funny), potentially to an uncomfortable degree:

P15: You have to be really ready to make a fool of yourself a lot of the time.
Interviewer: What do you mean?
P15: Just to be silly. To be a character. I think I take myself too seriously a lot of the time, and I think that hinders a lot of people wanting to listen to your music.
Interviewer: But ... you are who you are?
P15: Yeah, but I think everyone's kind of goofy, and it's just how much are you willing to show that, I guess.

Displays of the unguarded self are required to demonstrate intimacy, and being 'willing to show' that 'goofy' self has significant social (and commercial) benefits. Despite not answering directly to corporate imperatives – at least in this aspect of their lives – DIY practitioners felt restricted *because* of the intimate requirements of the platform. Online communication remains in an important sense highly *mediated*, insofar as the communication is substantially shaped and changed by the character of the medium. (It's worth remembering, though, that these platform 'requirements' are not an inherent, technologically determined aspect of social media platforms, but are changeable. They have developed over time in tandem with usage conventions – including within DIY specifically – as well as evolving platform business models, approaches to corporate governance, and state regulation, amongst other things.)

Importantly, the role of intimate authenticity on social media is a reversal of its political role as intended by DIY practitioners (and especially riot grrrls). Again, it relates to demystification. Social media teaches us that some people are 'naturals', and some people aren't. Rather than suggesting that 'anyone can do it', it suggests that certain people have innate communicative capabilities and rewards their ability to perform authentically.

Self-expression, representation and relatability

DIY has historically celebrated its capacity to give an outlet or 'voice' to experiences and identities that are not reflected in mainstream popular music. Early punk and post-punk scenes tended to be less concerned about representing any specific marginalized social group, and more with the failure of popular culture to represent everyday reality for its audience. As in post-punk band X's 'The Unheard Music' (1980), which decries 'some smooth chords/[...] no hard chords/on the car radio', links are made between the symbolic content of popular music and its role in the 'consciousness industry' (Enzensberger 1974). The argument, broadly, is that popular music in fact represents and reflects the ideology of society's most powerful, rather than the needs, desires and experiences of real 'ordinary' people. More recently, DIY has tended to emphasize the need to give voice not to generally 'absent' subjectivities, but rather to specific social groups. Riot grrrls sought to redress the representation of women on two fronts: as *mis*represented in popular culture and *under*-represented within the punk rock subculture which was supposed to present an alternative for those 'othered' by the mainstream. (In so doing, of course, they highlighted the ways in which the purportedly universalist concerns of earlier DIY movements were not so general at all.)

♪ 'Music's not for everybody' – Chain and The Gang

Whilst there is a meaningful difference between these two approaches, they share an interest in expanding the range of subjectivities that could be accessed through, and supported by, recorded music. And exponents of both approaches have tended to agree that the absence of any real commercial pressure is a part of what makes this 'expansion' of subjectivity particularly possible in DIY. This resonates with concepts developed by the sociologist Pierre Bourdieu, who saw a clear disparity between a DIY-esque style of 'autonomous production' and the realm of what he called 'heteronomous production' (i.e. that which takes place in commercial cultural industries). He saw culture's capacity for self-realization decreasing more or less in tandem with 'the increasingly greater interpenetration between the world of art and the world of money' (1995: 344). (A more populist perspective has sometimes inverted this argument, by seeing popular acclaim and the consequent economic freedom as the very thing that gives one the creative freedom to do what one wants.)

This interest in autonomous self-expression within DIY music cultures is concisely encapsulated in the notion of the 'no-audience underground', a term coined and popularized by Rob Hayler of the experimental music blog *Radio Free Midwich*.[4] This term, which has gained some critical attention – including a considered response from Simon Reynolds (2012) – suggests that DIY and left-field music might flourish in obscurity, primarily

---

[4]Hayler is Leeds-based and therefore has connections to the DIY scene covered here (although focuses generally on more free-form, experimental and electronic music). He has been quite clear that the 'no-audience underground' concept, whilst it has been engaged with academically, was not intended as a theory so much as an illuminating and amusing descriptor.

by relying on a high degree of involvement from a handful of people. For Hayler, part of the benefit of a 'no-audience underground' is a lack of artist–audience hierarchy. This is discussed in more depth in Chapter 4. Here, though, I want to focus on what the concept might imply for the relationship, in DIY music, between self-expression (i.e. speaking for one's self) and representation (i.e. speaking for others). Hayler writes:

> Some – most of the best – are compelled to create. The absence of standard recognition might grate occasionally but is largely irrelevant. These people do it because they have to or they love to or both. That someone other than themselves might appreciate their art is great, of course – none of us are without vanity-but not necessary. I know several people who, if shipwrecked on a desert island would be distracted from the business of survival by finding a shell that made an interesting noise when blown into.
>
> (Hayler 2015)

A 'no-audience' scene, then, might be beneficial insofar as a 'regular' audience might impose a kind of burden of representation on the artist. It relates to the idea that a radical (and resistant) aesthetics might *necessarily* be alienating and off-putting, and that an excessive focus on representation – that is, asking whether or not the audience might relate – is a kind of cultural populism on a micro-scale. That sense of freedom from audiences is neatly paraphrased by Czech music practitioner Jorge Boehringer as 'liberation through lack of interest' (2015).

A few of the Leeds DIY practitioners spoke about this sense of liberation. One older practitioner reflected on the value of DIY as a space where

> you can make selfish music, you know, music that is deliberately, not offensive, but difficult, challenging, but then maybe deliberately anti-social almost, as well. I think that's an interesting area, cos that's like 'I honestly don't care at all about what anyone thinks of this', that's a good thing to be doing with your time, you've got no rules to break. (P13)

Most practitioners though – particularly the younger ones – didn't seem to feel this kind of freedom. They wanted to express themselves – to give voice to their own emotions and experiences – but at the same time they wanted their output to serve an empathetic, representational function, by talking and singing about subjects that their audience (often largely friends) could relate to. So, while a 'liberation through lack of interest' framework usefully highlights the dangers of 'losing one's self' in the pursuit of popular acclaim and commercial success, it is important to note that self-expression and representation are not aspirations that inherently endanger each other.

Adam Arvidsson, in his work on the potential for an 'ethical economy' online, uses the phrase 'socially recognized self-realization' (2008: 332), which I think captures some of the ambivalence at work here. Arvidsson means it fairly positively, in terms of the way in which some activities (especially creative ones) are valued in part because they allow us not only to excel, but to be *seen exceling*. He sees this socially produced sense of being valued – which he connects to the ancient Greek notion of *philia* – as having the potential to form the basis of a new online economy.[5] But his phrasing also alludes to the more questionable aspects of what sociologist Anthony Giddens has called the 'reflexive project' of the self (1991: 53) – as in, the new burdens and responsibilities accompanying the distinctly modern task of working to create and perform one's own identity. In this framing, it is not a case of asking whether or not reflexivity enters our communicative and creative process – since it inevitably will, and indeed it is always-already embedded. There is no 'pure' pre-social self to express. We can, though, still seek to understand and assess the role played by social structures and institutions in augmenting and governing our reflexive self – which includes, in this instance, the role played by social media platforms.

♪ 'Only acting' – Kero Kero Bonito

Social media activity is often associated with corporate (and sometimes state) surveillance, but it is important to note that it also involves a huge effort of surveillance undertaken on each other (Andrejevic 2004). Indeed, the productive capacities of platforms depend not only on our communications with one another, but also on our ongoing (often quantified) *evaluations* of one another. Practitioners' output, whether musical or otherwise, is perennially put up for judgement online, and social media feedback serves as one of practitioners' main sources of validation. This is particularly true of Twitter, where retweets often signify an identification with the original post. Part of the decision to retweet is assessing the extent to which it might 'speak for' one's self; equally then, part of the process of composing tweets is asking one's self whether it might 'stand in' for the feelings of another, that is, 'will they relate?'

This anxiety over relatability can produce a 'chilling effect' on social media (Marder et al. 2016a), whereby users show a reduced capacity for self-expression when they are aware of the visibility of their activity. One Leeds DIY practitioner expressed a similar kind of wariness:

---

[5]Or at least, he did in 2008, when the existing internet economy looked quite different.

> There's always a thing with Twitter or Facebook that you're making your opinion public and you always wonder, or at least I do, what sort of motive is behind that, and it's not necessarily an ulterior motive … if you're gonna be like 'oh this band is brilliant' or whatever, as I say, there's always an exchange, you want some sort of reciprocation of your comment or whatever, and it's just, that seems … quite calculated a lot of the time. [...] You're essentially parading your morality, or what you perceive to be your good act to the world, and on a micro level I think that's what people are always doing with tweets and comments. (P17)

It's possible to read this as a damning judgement on the sycophancy of fellow practitioners. But the point, I think, is that platforms' communicative norms mean this kind of 'parading' is often unavoidable, even if it is not the primary intent. This stems in part from the distinctly promotional character of social media as it functions for most music practitioners. Even if you intend your live show to be a cacophonic, audience-repelling monument to solipsism, your social media content still needs to successfully encourage an audience to the venue in the first place.

Some practitioners felt surer of their ability to resist the impetus towards relatability and impression management, and thus to maintain a sense of authentic self-expression online. One practitioner recounted: 'Sometimes I'll tweet stuff and then I'll lose followers, and I'll think "oh, maybe it was that particular thing that I said." [...] But I don't find myself thinking "oh, I shouldn't say this cos it'll be unpopular"' (P9). In general though, as outlined in the previous section, practitioners felt pressured to adhere to a relatively constrained online persona, usually emphasizing humour and self-depreciation.

These issues of relatability and self-branding overlap with the broader issue of representation in DIY. As outlined above, the notion that scenes ought to provide a space for structurally disempowered and disadvantaged social groups has become steadily more central to DIY ethics over the past thirty years. There is widespread understanding that greater diversity in this regard might result in music that reflects more accurately the facets of life which are specific to certain groups (and which are erased in, say, more heteronormative, patriarchal, white culture). This is summarized in the maxim: 'you can't be what you can't see', a phrase which served as the title for a day of talks and presentations at DIY Space For London (DSFL) in 2016 (DIY Space For London 2016).

♫ 'Cool generator' – **Bad Moves**

That indie-punk DIY music has a serious problem with representational diversity is in no doubt. Indie and punk have historically tended to be articulated in ways that privilege the affective needs and creative output of white cisgender men, something that is widely acknowledged in both the academic literature (Bannister 2006b, Kruse 1993) and journalistic accounts (e.g. Sahim 2015). Social media has sometimes been presented as a space in which these hegemonic norms can be quietly bypassed or loudly undermined.[6] But while social media platforms bring new potential outlets for representation, it is important to note that online norms of relatability – while they create expectations and pressures for all of us – create additional pressures for those marginalized along classed, gendered and racialized lines.

Akane Kanai's research on meme culture and popular feminism is highly relevant in thinking about how an emphasis on 'being relatable' online might exacerbate existing processes of marginalization and exclusion. Her research shows that 'attaining relatability' online is premised on an ability to display a 'normative sameness' (2019: 142–50), and that certain relatability tropes like the 'confession of flaws' (very much akin to DIY's self-depreciation) have a complex reliance on whiteness and on middle-class status (2019: 47). (See also Pitcan, Marwick and boyd 2010, on the burden on people of color in the US to perform 'respectability politics' online, which tend to reflect 'neoliberal, White, bourgeois normativity'.) The particular normative dimensions of indie-punk DIY music may differ from those in Kanai's study, but the basic logic is the same. The social media pressures described in the previous section – to be funny, to be intimate, to be *real* – are not simply individual, psychological challenges; they also form the 'structure of feeling' that creates additional difficulty for any practitioners whose mode of self-expression clashes with hegemonic expectations within the scene.

Communication in DIY music, whether through live music, recordings, zines, or social media content, is never devoid of structuring elements. Even in a seemingly 'no-audience underground', it is important to recognize that the DIY audience (malleable as it may be) does play this kind of normative, structuring role. Practitioners attempting to open up DIY scenes to a range of marginalized identities have recognized that reconstituting DIY audiences is a necessary precondition of fuller self-expression. For example, at the inaugural Decolonise Fest in 2017 – a London music festival organized by the DIY Diaspora Punx collective – the famous riot grrrl call of 'girls to the front' was powerfully reworked as 'people of colour to the front' (Phillips 2017). Such a spatial reorganization fully acknowledges that self-expression and representation might sometimes *require each other*, and that neither artist nor audience can function as an 'island' in this regard.

---

[6]Sasha Geffen suggests, in a similar vein, that online platforms played this kind of liberating role for an emerging trans music culture: 'The proliferation of internet-equipped consumer electronics enabled a new generation of gender nonconformists to communicate across any distance. Trans kids no longer had to move to New York or San Francisco to speak with others like them; they could use Facebook, Twitter, Tumblr and YouTube to find community' (2020).

However, a logic of relatability seems to largely act against any equivalent spatial reorganization on social media, by denying that the audience themselves might need to take some shared responsibility for any individual's capacity for self-expression. As Kanai's work shows, the imperative to 'relate' is an imperative to play the room 'as its dealt', and to take on responsibility (i.e. to blame one's self) for any communicative failure. The delicate, imperfect potential of 'socially-recognized self-realization' is thus severely imbalanced. If we seek to enable, in Nancy Fraser's terms, 'recognition that can accommodate the full complexity of social identities' (2000: 116), it is necessary to pay close attention to how the celebration of relatability places this burden on individual practitioners, and on marginalized and oppressed individuals in particular, to self-govern themselves into normative subjects under conditions of surveillance and scrutiny.

Everyday social media use and the DIY 'audience commodity'

Being a social media user, even as a music practitioner, always involves more than just 'producing' content. DIY practitioners are the kind of social media users that Jose van Dijck describes as 'both content providers and data providers', which is to say that although they use platforms to produce, this does not prevent them from also being profiled by platforms as a consumer and having their data collected, utilized and returned back to them in the form of targeted advertising (2009: 47). All users, says van Dijck, 'whether active creators or passive spectators [...] form an attractive demographic to advertisers' (2009: 47). In short, DIY practitioners are, at least sometimes, an audience, and they are surveilled as such by platforms eager to pin down their taste profile in order to improve advertising efficiency (Andrejevic 2007). Time spent uploading music or promoting a show is insubstantial in comparison to the amount of time that most practitioners – including myself – spend scrolling passively through news feeds, disinterestedly clicking links or half-watching video content.

DIY music practitioners certainly do not imbue this activity with the political meaning that they tend to attach to playing shows, making music, or the work of forming and building musical communities on- and offline. Arguably, therefore, it falls outside of the remit of this study of DIY activity. But practitioners' political perspectives on social media platforms are just as likely to be shaped by their everyday experience of 'living' with (and on) them as by their specifically musical online activity. And given that DIY has so often been about, in a broad sense, 'overturning' consumerism in favour of production, it seems highly pertinent that platforms might seem to collapse distinctions between these two modes – an issue I return to in this chapter's conclusion. This section focuses primarily on Facebook, as the platform that most often came to mind for practitioners when discussing these issues.

One relevant concept for considering the political dimensions of everyday social media consumption is via Dallas Smythe's influential notion of the 'audience commodity' (1977). This concept recasts media consumers (in Smythe's original study, television viewers) as 'the principal product of the commercial mass media in monopoly capitalism' (1981: 26). That is to say that, in the media economy, the 'product' is often the viewing (or listening) time of audiences, sold to advertisers by media firms who compensate viewers with the 'free lunch' of media content. Thoroughly critiqued and debated at the time of its first description (Jhally 1982, Livant 1978, Murdock 1978, Smythe 1978), this notion has been reconfigured in recent years to consider the ambiguous 'work' of social media participation as a similar process in which users (or, their time and attention) are the products sold by platforms to advertisers (Fuchs 2012, Lee 2011, Manzerolle 2010, Manzerolle and McGuigan 2014). Critics of this application of the audience commodity thesis argue (amongst other things) that it suggests a 'social factory' extended *ad absurdum*, to the point where all activity, whether paid work or leisure time, is considered exploited labour (Caraway 2011, 2016). In the digital age, the debate continues as to whether social media activity constitutes 'free labor' (Terranova 2000) or whether, in characterizing such activity as exploited work, we deny users' subjectivity and potentially restrict our ability to meaningfully criticize actual (or, at least, *worse*) exploitation of workers (Hesmondhalgh 2010).

However, here I want to temporarily sidestep issues of what does or does not constitute unpaid work or exploitation, in favour of considering the subjective, psychological dimensions that might come with being *aware* of one's status as an 'audience commodity'. In particular, my assessment centres on what has generally been considered the most significant subjective consequence of commodification: alienation. Marx sees alienation as a consequence of humans' labour being redirected to ends beyond their control (since ownership of the means of production enables capitalists to harness labour power in this way). Separation from one's own labour extends out to wider alienation: from one's self, from one's colleagues and from other people in general. Robert Blauner's empirical research adds some useful specificity to this concept which can feel ungraspably abstract, suggesting four categories of workplace alienation: powerlessness, meaninglessness, isolation and self-estrangement (1964). Whilst Blauner was concerned primarily with industrial production, rather than the digital consumption that is the focus of this section, these categories might nonetheless illuminate the subjective dimensions of being part of the platform 'audience commodity' today.

DIY practitioners in Leeds were largely aware that Facebook was interested primarily in capturing their data and selling that data to advertisers. But, like most of us, they found it hard to express how that specifically connected to or altered their approach to using the platform. For example, when asked how they felt about Facebook as a corporation, one interviewee offered the following: 'We are the product, and they are selling us to companies as people to consume adverts'.

It is hard to imagine a more concise summary of the audience commodity thesis. But the same practitioner continued:

> I know it but I don't care enough cos it just literally like, without Facebook there's a lot of things I wouldn't know about. [...] I guess I'm willing to sell a little bit of myself to be able to use Facebook as a means to find out when gigs are. (P4)

Clearly, this experience of using social media – with all the benefits of participation and communication on offer – is a far cry from the kind of self-estrangement identified by Blauner, which occurs when work offers no opportunity for self-expression. Neither does this experience seem to necessarily alienate users from other users; indeed, an increased social connection is the *purpose* of the activity (although the quantified competitive aspects of these platforms have negative effects that are addressed elsewhere).

In terms of identifying their commodification as an audience, practitioners were naturally delimited by their standpoint. It can be difficult to see Facebook from a perspective other than how it feels to use it. But they were also limited by the amorphous nature of the platform and by the difficulty of understanding how their social activity translates into a business model: 'It's kind of difficult to think of it as a corporation because so many people are on it that it becomes sort of like ...' (P3).

The unfinished sentence here is appropriate. What *does* it become? This practitioner is not alone in struggling to find language appropriate to describe the tangled web of sociality that constitutes our subjective understanding of Facebook as an entity. Facebook is such a powerful objectification of social relations that in some sense it *is* those relations.

Autonomous Marxists have posited the 'general intellect' as the great hope of an egalitarian technological future, as immaterial labour is collectivized and no longer works for capitalist ends. Nick Dyer-Witheford, writing in this tradition at the end of the twentieth century, proposed that the internet offered enormous potential for reorganization of work and society through the emancipation of this general intellect, even as capitalism made great strides to bring it under its control (1999). On Facebook, it is precisely this general intellect that we are estranged from – the weight of social relations, objectified in its representation *as* Facebook, appears immovable.

Big platforms like Facebook also carry a sense of being ubiquitous and omnipresent, as humorously identified by two Leeds practitioners in a group interview:

P5: I think in the early days of Facebook, Mark Zuckerberg said he wanted Facebook to be as common as turning on a light, and I guess it's become that.

P4: It's more common than turning on the light! You can go on your laptop in the dark.

One practitioner spoke of the difficulty of having a 'political stance' towards social media 'when it's so omnipresent with everyone you know that it's impossible' (P3). The platforms' power over the general intellect is such that one's value (or labour power) feels scarcely worth withdrawing.

For those who identified social media as an antagonist, one real challenge was in finding a way to think and talk appropriately about it:

> It's hard to formulate an actual counter-stance without sounding like some tin foil hat wearing conspiracy theorist. But yeah, I am worried about my data, that self online being monetized, being abused, in a way. But, at the same time, I chose to enter the data and that's how the system propels itself. I mean to be honest it's not something I think about an awful lot, because you can't think it too much, you can't think about what being on the internet and doing work on the internet … you can't. (P17)

More than one practitioner identified themselves as having a 'paranoid' perspective on social media platforms and data. Such a perspective is perhaps not inappropriate. One longitudinal study of Facebook users in the United States showed that even as concern over maintaining privacy grew, and users attempted to reduce the amount of personal information they shared, they were unable to prevent an *increase* in the information they divulged to data-gathering 'silent listeners' – Facebook, third-party apps and advertisers (Stutzman et al. 2012). DIY practitioners knew that data was being produced and captured, but they weren't sure what, or when, or where it went, and they weren't always sure why.

Practitioners often rationalized their relationship to Facebook through reference to the status quo, drawing on norms of free market exchange to justify or explain their tolerance of the platform:

> They're a business and we're using them. It's not like promotion for your art is a human right, is it, really? […] The people that program Facebook are professionals with skills that have created this revolutionary tool and they should generate income. That's how capitalism works, and sadly we're all part of it, whether you like it or not. (P14)

There was a widespread view of Facebook as relatively fair *within a capitalist system*. Even for those who felt that Facebook wasn't exactly a great corporation, it was a struggle to identify the specific things that it was doing wrong[7] or to find a reason to proclaim it as worse than any other big firm.

---

[7] The majority of my interviews took place before media coverage of Facebook's role in the Cambridge Analytica data scandal reached its apex in 2018 (Cadwalladr and Graham-Harrison 2018). Public fears over online disinformation and polarization were at this point only beginning to coalesce.

So, DIY practitioners didn't generally see themselves as suffering at the hands of social media platforms, even when they were able to accurately identify the means by which the platform generated surplus value through their activities. This is in part because alienation is experienced in tandem with (and often to a lesser extent than) empowerment. Returning to Blauner's categories of alienation, there is very little evidence of 'self-estrangement', since practitioners had plenty of capacity for self-expression online, but they did have feelings of 'meaninglessness' and 'powerlessness' which existed alongside the meanings that they made for themselves. Certain types of alienation, then, were notable primarily by their absence – one of the fundamental differences between social media activity and most paid employment is in its high level of autonomy and self-expression. Practitioners rarely feel truly *compelled* to be active on Facebook and, even when they do, it is social relations rather than Facebook as a corporation that compels them. It would be inaccurate to consider them 'estranged' from their work in the sense that Marx and Blauner identify, because they have control over the content that they post. But they also know that something else is happening, even if they can't say exactly what. The value of what PJ Rey calls 'ambient production' (2012: 410) – the quietly captured data that, as my interviewee identifies above, 'propels' the system – is undoubtedly alienated from practitioners.

In Marx's vivid, evocative descriptions of alienation, the products of labour return to loom large over their creators in hostile and unrecognizable forms, masking the exploitation that occurs through the capitalist appropriation of their surplus value (Marx 2000: 322–33). The commodification of audiences on social media is a murkier process; it is the indistinct shadow cast by the brashly illuminated activity of social media participation. From this darkness comes the pervasive and nagging voice which murmurs to practitioners that their agency is undermined, even whilst the visible world of social media brightly affirms that it is not.

In the remainder of this section I consider the (mostly minimal) means by which practitioners do attempt to resist experiences of commodification and alienation as users of social media. Although I present them here in terms of resistance, it is important to understand the extent to which these practices are considered to be personal preferences rather than political praxis, and also the extent to which these practitioners feel ambivalent towards Facebook, and are very willing to consider its positive effects. Even if practitioners dislike aspects of Facebook, or distrust it, it is rarely considered to be their primary antagonist or opponent. The dynamic of corporate versus anti-corporate is acknowledged but it is by no means the only dynamic in operation. Facebook is something that practitioners work with begrudgingly, that they sometimes jostle up against uncomfortably, that sometimes is a valuable tool, that sometimes feels like home and sometimes feels as though it isn't there at all. Therefore, when considering the steps that DIY practitioners take to deflect or mitigate platforms' capture of their activity, it is important to understand these actions as a part of a wider range of practices – not all of which are performed in the name of resistance.

One straightforward way to minimize one's data capture is to leave Facebook or, even better, to eschew signing up in the first place. A collective 'exodus' of social media wasn't seen as feasible amongst DIY practitioners in Leeds (although there was an interest in potentially amenable alternative platforms, such as the briefly touted 'creators' network' Ello), but individual non-participation through deleting or deactivating accounts, or through ceasing to post, was reasonably common. This was often considered as a practice undertaken for one's sanity or well-being, rather than a political strategy, and even those doing the deleting and deactivating were sceptical of its value in this regard:

> Interviewer: Why don't you think it's resistance?
> P17: Well I think it is resistance, but it's a very mild form … I don't want to overstate the implications of me deleting all my posts from Facebook and making my photos private. I really don't know. But at the same time, I feel like it's become such a part of everybody's lives that when someone like [my friend] deletes his Facebook, everyone laughs, and thinks 'oh you know, that's such a pose', it's such a 'look at me' statement. I don't know, it's kind of complicated, but I'm not sure I'm being radical in any way, or even political.

Non-participation, then, is all too easily read as an expression of superiority, and a way to distinguish one's self from the crowd. Deleting or deactivating social media can also carry associations of poor mental health, and digital non-participation is considered to be a useful period of respite necessary in order to 'recharge one's batteries' and return rejuvenated. Such a perspective reflects the extent to which opting out of social media means opting out of social life – that is, it must only be left temporarily.

Another common approach was a policy, however informal, of minimal use. This is summarized in the phrase 'I don't really use it, apart from … ', where usage is usually restricted to finding out about (and promoting) shows, asking for favours, or crowd-sourcing recommendations. This was not really identified as a strategy of resistance, but practitioners did recognize that this made them at least a 'non-ideal' user from Facebook's perspective. A more consciously resistant approach was to knowingly provide incorrect information on social media, in particular through the use of pseudonyms:

> I think a lot of people kind of implicitly reject this idea of themselves as a product on Facebook ... I guess by the way they present themselves, people who won't have their actual name on Facebook. People having like comedy names and things, plays on their real name. (P5)

Obfuscation of this kind threatens the accuracy, and therefore the value, of Facebook's primary product: data (see Casemajor et al. 2015: 861). Just as the audience commodity is a means by which TV networks attempt to package viewers as a product in order to sell them by the unit, social media behavioural data is designed to demonstrate to advertisers that a platform's knowledge of consumers is suitably accurate and worth paying for. It also serves to ridicule, however faintly, the idea that the real self and the Facebook profile are one and the same, and the idea that capturing data is meaningfully equivalent to capturing the person generating it. There have been attempts to counter this obfuscation, most notably Facebook's 'real names' policy. This was met with widespread criticism, particularly from transgender users who no longer went by their birth name. They argued that this was a transphobic policy which in some cases would compromise users' safety (Holpuch 2015). Facebook retreated on this issue, suggesting that, for all its power, it does rely on active, *willing* users, and a degree of obfuscation may be part of the price the platform pays for this.

Some practitioners also understood themselves as resistant social media users insofar as their DIY music activity made them, in a sense, 'bad' consumers: 'I think seeing as the majority of my Facebook likes are for DIY bands, Facebook probably finds it quite hard to market small bands with like several hundred likes ... it probably finds it quite hard to make those sorts of adverts' (P5). DIY music here is seen as a niche market which Facebook's algorithms might lack the nuance to infiltrate. There was also a general sense, especially amongst younger practitioners, that they might be a little too culturally and politically savvy to fall for the blunt instruments of Facebook marketing.

However, the main way in which practitioners saw themselves as resisting social media norms was not through any consumption strategy, but through the specifically political nature of their productive DIY activity. One practitioner remarked: 'It's ironic that something so corporate [i.e. Facebook] would be a platform for something so anti-corporate.' This connects the production discussed earlier in the chapter to the consumption discussed here – the former served to nullify the latter for some practitioners. However, as I have identified, there was also widespread, low-level unease and insecurity about the extent to which they were commodified through data capture. The sense of DIY music as separate from systems of commodification, which may still be clearly distinguished in its production, is undermined by the commodification of the everyday activity within the scene.

Production, consumption and self-realization

Whilst this chapter has focused on DIY practitioners' subjective experiences of social media, it has also attempted to articulate the different kinds of subjective experiences offered by production and consumption. Although production and consumption would appear to be two sides of the same coin, with each instance of one creating an instance of the other, they tend to be imbued with very different moral valences. The anthropologist Daniel Miller has identified a longstanding (i.e. pre-industrial, pre-capitalist) ideological tendency for societies to view production as the 'creative [...] manufacture of value', and consumption as 'the using up of resources and their elimination from the world' (Miller 2001: 2). Consumption has, across a wide array of societies, cultures and religions, been denigrated as the wasteful antithesis to production, as vulgar and excessive or, in a moral framework Miller links back to Eastern religions, as 'the wasting away of the essence of humanity in mere materialism'. 'This makes it quite unsurprising', he continues, 'that the earliest discussions about consumption which were written prior to the rise of capitalism look remarkably similar to contemporary discussions' (Miller 2001: 2–6).

DIY has historically offered a critique of consumerism which proposed cultural production as part of the solution: a source of autonomy and empowerment in opposition to the comparative restriction of consuming music commodities. This has been called into question by contemporary practitioners, who are increasingly attuned to the benefits that cultural consumption offers. But producing culture – making music – does continue to engender quite special affective states, such as the feeling of crossing a significant threshold of self-realization.

Social media has brought about new opportunities for practitioners to produce (and consume). But as this chapter has shown, the forms of production that are most empowering tend to remain tied to directly socio-*musical* processes, rather than more general online activity. Less rewarding forms of production often involve feelings of compulsion, alienation and anxiety. What DIY hasn't yet developed is a critique acknowledging that a *compulsion* to produce culture might be, in some circumstances, just as restrictive as a compulsion to consume it – i.e., that consumerism is paralleled by what we might awkwardly label 'producerism'. This is particularly pertinent in online spaces, where logics of participation can compel us to 'take part' or to 'speak up' in ways that can feel momentarily empowering, but which may not meaningfully link with any particular democratizing or emancipatory political project (see Barney 2010). Consumption itself is not inherently pernicious (indeed, it is necessary and unavoidable), but is made so by a capitalist ideology of consumption which overstates the correlation between the products we own (or would like to own) and the way we would wish our lives to be. Cultural production can obfuscate and overstate in similar ways and can accordingly be an obstacle to, rather than an apparatus for, self-realization.

4

The players

Hierarchy, ownership and collectivism in DIY scenes

Although DIY stands for 'doing it yourself', the practitioners I interviewed in the Leeds scene often emphasized that DIY music was really all about 'doing it *together*'. Togetherness – whether considered in terms of solidarity, cooperation or collective participation – is a key means by which DIY scenes understand themselves as culturally resistant, and as distinct from the more competitive, individualist world of popular music. Having considered social media's impact on self-presentation in the previous chapter, this chapter looks at what DIY practitioners do together and assesses how social media has affected the quality and quantity of cooperative feeling within the scene. DIY's 'outward' relationship to other scenes, and to the rest of the world, is considered in the two following chapters.

DIY's 'anyone can do it' call-to-action is implicitly anti-hierarchical. Pop music tends to orbit around the cultural impact of superstar performers – partly in an effort to mitigate the inherent unpredictability of the music business (Toynbee 2000: 16–17) – creating a concentration of wealth, power and attention around a handful of best-selling artists. Part of the claim to cultural resistance made by DIY music scenes has been that their social composition is 'flatter' (i.e. less hierarchical) with fewer distinctions between 'big' artists and small, and also less separation between performers and the 'ordinary' members of the audience (Dale 2012: 6). Practitioners I interviewed in Leeds often emphasized 'participation' as a critical goal, and there remains a strong interest in dismantling artist–audience hierarchies and critically interrogating power dynamics within the scene.

The internet has been affiliated, in a variety of ways, with similarly anti-hierarchical, 'anyone can do it' ethics. The internet's capacity to bypass existing cultural intermediaries and gatekeepers was seen as bringing about new possibilities for deserving 'ordinary' musicians to become stars, especially in the early 2000s (although, in truth, most of these famous 'MySpace success stories' were built upon hidden contributions from 'old' intermediaries such as managers, labels and PR firms). In this period, it was also seen as posing a substantial threat to the liberal intellectual property regimes upon which the recording and publishing industries relied – partly through so-called 'piracy' (Prior 2015) and also through a burgeoning (but ultimately limited) 'remix culture' associated with early 'Web 2.0' platforms (Howard–Spink 2004). More broadly the internet was seen as allowing new opportunities for 'peer-production' and for new, ambitious kinds of collective ownership (Bauwens 2005, Bruns 2007, 2008a).

Many would argue that hopes for a collaborative, 'commons'-style internet have been usurped by the monopolistic giants of platform capitalism, but this doesn't necessarily eradicate the internet's redistributive potential entirely. Fuchs and Dyer-Witheford have argued for the need to understand the internet in terms of dialectics – existing modes of production 'anticipate' their critique, and therefore we can 'see' the ways in which current tools might, in different hands, allow for the radical reorganization of society (2013: 786–7). I have suggested that DIY music should not be considered as this kind of radical political praxis, but DIY activity does nonetheless give some insight into what some 'different hands' might do with currently dominant platforms.

In the first section of this chapter I consider DIY's capacity for collectivism and cooperation. DIY does have a 'status order' inherited from popular music which results in hierarchies, but this status order has some specific characteristics that might still valuably support attempts towards social justice. I suggest that a scene with such internal hierarchies might nonetheless find value (and take pleasure) in each other's achievements, using Harvie's notion of 'convivial competition' (2004). This conviviality is still present online, particularly in practitioners' engagements with social media algorithms, but 'networked' engagement with DIY activity does diminish feelings of collectivism and solidarity.

The second section considers how old forms of DIY ownership – which have sometimes gestured towards collectivism – have impacted upon new social media offerings. Pre-social media online forums were effectively an extension of existing independent distribution networks, but their contemporary equivalent, Facebook Groups, quickly devolve into self-interested promotional mechanisms. Facebook Pages is a much more atomized and individualized means of enclosing ownership over activity, but are the norm, both inside DIY and in wider popular music. Some DIY practitioners engage in more radical responses by emphasizing anonymity and through attempts at creating an online 'commons'. I conclude the chapter by arguing that it is social media's capacity to individualize and commodify social processes that constitutes the most substantial threat to collectivism within the DIY scene.

Collectivism and 'convivial competition' in DIY music

As I outlined in Chapter 2, DIY music has a long history of operating 'not-for-profit'. In some DIY scenes this has been seen as key to developing and protecting uncompromising aesthetics. But amongst my interviewees, the importance of a not-for-profit approach was understood as relating to an aim of encouraging broader and fairer participation. Distance from the market was understood to broadly correlate to distance from unhealthily competitive logics. DIY was, for one young practitioner, about 'like, the whole not-profit thing [...] not making like a hierarchy out of it, like nobody's earning loads of money, and then just like a cut of it goes to the bands or whatever. It's just like everyone trying to cover the costs of it, not making a profit' (P7). Some practitioners were able to identify what *wasn't* DIY by identifying the places and people who seemed to be more interested in the money than the music and the people. For example, one promoter recalled being 'marched to the cashpoint' by a punk band who clearly didn't share the same understanding of what constituted acceptable DIY practice (P19).

The not-for-profit ethos is intended to remove the economic incentive to behave self-interestedly. But an absence of profit does not mean a total absence of hierarchy. In considering how social status is accrued in DIY music, I draw gratefully on Matt Stahl's rich ethnographic study of the San Francisco indie-rock scene throughout this section (2003). Stahl argues that while the scene he studied often emphasized its inclusivity to musicians and non-musicians alike, it was in fact rife with 'processes of hierarchization':

> Jerry-built stages and light and sound systems in bars, cafes, small clubs, and even house parties elevate, illuminate, and amplify performers over nonperformers. In addition to money payments (however small), free drinks, 'backstage' areas, and guest lists privilege musicians over audience members. Local weeklies and zines run reviews, photos, interviews; college and community radio DJs plug local musicians' shows and feature their music and voices as interviewees and guest DJs, increasing their visibility and audibility in the local urban environment.

(Stahl 2003: 140)

Stahl suggests that the hierarchies he found are perhaps an inevitable consequence of the artist–audience separation within rock (and popular) music: 'In rock culture, sacralization of individuals and bands – the valorization of certain "non-economic" forms of capital they hold – is typically a unidirectional process, and no amount of "you've been a great audience" can change that' (2003: 157). Some scholars in the field of 'fan studies' might disagree about quite how unidirectional this artist–audience communication process is. But Stahl's critique is a pertinent one for DIY. DIY's not-for-profit ethic makes it relatively immune to some critiques of 'indie' and 'alternative' music as, essentially, petty capitalism (e.g. Keightley 2001: 129), but it is certainly not free of processes of social and cultural distinction.

Take the example of DIY promoters. Most would never dream of earning money from promoting a show, even though it requires lot of time and effort; DIY norms dictate that all revenue (after venue hire and other expenses) should go to the performers. But Stahl's description of the 'paradoxical' nature of hosting – 'at the same time servant and honoree, fulfilling and incurring obligation, deepening social ties' (Stahl 2003: 153) – is useful in considering DIY promotion as both a difficult, thankless task, and one that offers credibility and status. Even if this social status is not considered an appropriate motive – one interviewee argued that, as well as being not-for-profit, DIY ought also to involve 'not doing it for kudos and your own elevation' (P12) – the unavoidable fallout is what Nancy Fraser might call a 'status order'. This is the term Fraser uses to consider the non-economic forms of disparity that arise from 'socially entrenched patterns of cultural value, of culturally defined categories of social actors [...] each distinguished by the relative honour, prestige and esteem it enjoys vis-à-vis the others' (2000: 117).

There is certainly a 'status order' in DIY music. As Stahl suggests, and as I have covered in Chapter 2, it arises in part as a consequence of an adherence to the forms and rituals of rock and pop music. There are musical role-models who others seek to emulate, or who seem to characterize the aesthetic of the scene as a whole, and receiving recognition from these role-models can constitute a conferral of status. There are also cultural gatekeepers: people who have the social connections to bring people together and who can grant access to specific opportunities. Other studies of DIY music scenes have noted that it is these gatekeeping roles where a disparity of influence is most strongly felt (see, for example, Mullaney 2007).

DIY music has its hierarchies, like other music scenes do. But it is also distinctive, and it is important not to relativize this distinctiveness away to nothing: DIY does not have the same status order as the popular music industries. Efforts are frequently made to share resources and knowledge that would, in other music worlds, be hoarded. In the UK DIY scene, there are events like 'First Timers' at DSFL (DIY Space For London), in which only musicians or bands performing for the first time can play – the organizers describe the event as a 'celebration of demystification' (First Timers 2017). For technical skills such as sound engineering there is often an informal, not-for-profit 'shadowing' system, where asking an existing sound-person to learn how to 'do sound' constitutes a kind of *ad hoc* apprenticeship. DIY is also sceptical of some conventional rock tropes that seem to engender disparity, such as the notion of a 'headline act' at live shows – the last band to take the stage at DIY shows will usually be the band which has travelled furthest to play, regardless of their perceived popularity.

The status order in DIY is further distinguished by the valorization of some key practitioners for reasons unrelated to fame or fortune. There are 'moral compass' figures in many DIY scenes who serve as embodiments of good DIY practice, and whose activities are seen as offering an insight into what constitutes being (or not being) DIY. These figures are often recognized for their longevity (i.e. being a 'lifer') and can therefore be a source of important knowledge about the scene's history. They might have specific skills which are highly valued but which do not give them gatekeeper status – for example, guitar repairs, print-making or technical ICT skills. Whilst these people might also be cultural gatekeepers or musical role-models, often their relative obscurity on those fronts can work to demonstrate their own moral fastidiousness: one interviewee emphasized the virtue of practitioners who 'toil away for years' without recognition (P14).

I have made the argument that DIY's status order is distinct in some important ways, but that it also falls quite far from its own discursive ideal. Without wanting to essentialize social processes down to 'human nature', it does seem somewhat inevitable that within current DIY structures, some bands will be more or less popular, that some will end up more central to the scene than others and so on. This suggests that, in seeking to assess and advance social justice within DIY's own borders, we might need to think in terms of something more immediately achievable than a full realization of what Nancy Fraser calls 'participatory parity' (2000). Instead, I suggest that the aims of DIY music might be more fairly assessed in terms of 'convivial competition'. This is a term that Marxist economist David Harvie employs to consider competition and collaboration in academic scholarship, arguing that competition can be 'convivial' when there is a broadly shared objective, when it is sufficiently distant from the concerns of the marketplace and when it is helping to motivate people to contribute to a 'commons' (2004: 4). This kind of competition, Harvie argues, can be arranged in such a way as to bring benefits to the community as a whole.

♬ 'My rights versus yours' – **The New Pornographers**

This notion, that camaraderie and solidarity needn't inherently be prevented by the presence of a competitive dynamic, resonates with the organization of small music scenes more generally, where individual goals are often operating alongside shared aspirations. In her classic study of music communities in 1980s Milton Keynes, Ruth Finnegan found that this sense of conviviality was common: brass band members identified strongly with their own bands, but also with 'the "brass band movement" as a whole'; folk musicians similarly had a 'strongly held, if not always articulated, set of ideas about the kind of enterprise in which they were engaged' (Finnegan 1988: 52–66). The brass band example is particularly germane since it is a world of musical activity organized specifically around competitions, in which bands are pitted against one another for prizes and accolades.

In DIY, collective production is linked to some extent to ideas of shared values (like the brass band movement), but it also draws upon its political alterity and sense of cultural resistance. This provides an often-powerful conception of DIY as a communal project in which all participants are seeking approximately the same ends. This feeling is well illustrated by the following excerpt, in which a practitioner outlines the way they felt the DIY scene was composed of people with a similar outlook:

> It's always felt like people are just happy that there is interesting music that's being created, and people want to share it, and if you like something, you want to do your best to make sure as many people as possible can hear it. Even if, and probably particularly if, it's not your own stuff. If you hear a fantastic track by [DIY band] or whoever, you think that's a brilliant track, and you want as many people to be able to hear that. So from that perspective, part of it is wanting the best for these artists and those acts because you're in a similar position and you'd think 'if they thought that about my track, they'd do the same'. And that's not necessarily competition, it's that we want the world to be filled with brilliant music, rather than crap music, so as many people as possible should be hearing, you know, that song or that song or that song. (P22)

This highlights the extent to which reciprocity is fundamental to any sense of conviviality in the scene (i.e. 'if they thought that about my track, they'd do the same'). A similar kind of reciprocity is identified by Matt Stahl, who argues that the San Francisco indie-rock scene is based on assumptions that demands made for attention and status will be paid back: '"You may not be in the spotlight this time", goes this logic, "but your attention to another who has claimed the spotlight guarantees you a brief tenure in that position in the near future"' (2003: 158).

But whereas Stahl seems to see this expectation of reciprocity as evidence of a harmfully self-serving social logic in his scene, I don't think reciprocity is inherently individualistic. It can be the basis for expressions of solidarity and empathy. Here, for example, two interviewees overlap in their consideration of what defines DIY participation:

> P24: I think it is literally about supporting other people who are doing similar stuff to you and as a result, because you do that, you know or you hope that you're gonna get that in return. Yeah, that's maybe how I view it, and that's certainly why now I always make the effort if I know people … even if I'm not a million per cent into a band a promoter is putting on. […] And I think you don't do that cos you're like 'well I'd better buy this, cos they'll buy that', but you do it cos you want to make sure that they can continue to do that …
>
> P23: … to do what they care about. And then you get into things through doing that, through having that open mindedness. And it's not like an in-crowd loyalty, it's like a … I guess it's like a co-operative attitude. You end up being exposed to things by having each other's backs.

The idea of 'having each other's backs' is quite different from a cynically minded trade-off and seems to emphasize the value of sparing each other from unsuccessful shows. The potential DIY audience is largely made up of DIY practitioners, all of whom will know the unhappy feeling of a near-empty venue.

Of course, there are sometimes feelings of resentment. Conviviality is diminished when practitioners feel like their efforts to support others aren't being adequately reciprocated. One interviewee reflected, regarding other scene members:

> I can't make them come to shows. It's their life, I can't make them feel guilty for not supporting me cos at the end of the day it's a charitable thing and I can't feel bitter at people if they don't come to shows 'cos it's the luck of the draw or whatever. I try not to think like that cos it makes you resentful. (P1)

Even here though, there is an awareness that being resentful might be an inappropriately individualistic way to respond to an unequal distribution of status, time or attention.

The extent to which the internet has been considered a competitive or cooperative space has varied substantially across its history. At the time of writing, however, 'convivial' is probably not the first word that would spring to mind to describe the tone of most online discourse. And social media in particular has often been presented as a place where the presence of so much cultural content, made freely available, creates an 'attention economy' characterized by intense competition for views, like and plays (Boyd 2017,

Goldhaber 1997, Marwick 2015). But DIY music scenes online do make attempts to retain a sense of collectivism, often based on the principles of reciprocity outlined above. This can sometimes mean trying to keep practitioners grouped together, across an online environment that is ever-changing:

> P28: A lot of us who initially started using Facebook more, it's like, 'we just have to click Like on each other so we all stay connected'.
> Interviewer: So that's because of the Facebook algorithms?
> P28: Yeah. But this is because these are all the people who were the carryovers from MySpace, this is like, 'how do we make sure that we keep each other in the same news feed, that we're all still showing up?' … and that's how. If you wanna promote someone's music, click Like. Even if you don't like it, click Like if you think someone you know should know that that exists. […] You want people to know it's part of your world.

In taking this collective action in order to 'keep each other in the same news feed', this group of practitioners based their activity on an understanding (or assumption) of how it will be interpreted and utilized by social media algorithms. This lending of one's own social status is akin to 'signal boosting', a technique which has been used to promote a diversity of voices in feminist blogging and in online circles concerned with intersectional social justice more generally (Rentschler 2017). This is important in the context of recent literature in social media studies, where a great deal of attention has been paid to how people – and cultural producers especially – attempt to understand and manipulate algorithms (Bishop 2018, Cotter 2019, Petre, Duffy and Hund 2019). In such accounts, this 'gaming' or 'hacking' of algorithms is almost always assumed to be to utilized in order to gain more attention for one's self. The 'gaming' of algorithms here, though, takes place as attempt to maintain the kind of reciprocity that I have identified as a common-place motivation within the DIY scene.

On the other hand, there are some specific features of the social media landscape that do threaten feelings of collectivism in DIY. Convivial competition relies on a sense of belonging, which platform architectures can often negate in favour of what social media research danah boyd calls 'networked publics', which 'force everyday people to contend with environments in which contexts are regularly colliding' (2011: 47). In this way, social media platforms can reconfigure or blur boundaries between DIY and other scenes. The sense of 'shared goals', for example, is compromised by features like SoundCloud's 'autoplay', the algorithms of which seem to make no attempt to follow a user's choice of track with something aesthetically or even geographically related:

If you click on somebody else's track [i.e. someone in the scene] and you listen to it, suddenly you find yourself listening to something completely different, cos you let it cycle through and it's 'other tracks you might like', or something. Like oh, what's this, I've never heard of it, and that's not what I expected to hear – I'm not really in the mood to listen to some electro-soul at the moment. (P22)

Platforms, and their 'networked publics' approach, have a tendency to posit and test new connections between actors, which can sometimes feel jarring. This might be considered as an accidental consequence of platforms' business models, where data gathered on connections made and unmade between 'nodes' of social networks is the source of behavioural profiling. But it can also be seen less deterministically as a consequence of the 'connectionist politics' that have been a major influence on Silicon Valley discourse (Turner 2017). Fred Turner's study of the intersections between early internet culture and radical sixties counterculture shows how the utopic vision of connecting all people, allowing for global communication outside of state apparatus, was a key normative principle of 'cyberspace' (2010). For better or worse, DIY scenes have historically tended to be sceptical of this kind of connectionist ideal and the universalist approach to music that online networks can imply.

Even when only considering content from inside the DIY scene, network dynamics are still at work in showing or making national and international connections that would otherwise be beyond the scope of local practitioners. The sheer amount of material online can be threatening or off-putting, even when one acknowledges that there are shared aims and tastes which might form the basis of an understanding:

P23: Like, sometimes I'll be having one of those days when you're looking for new music all day [online], and I'll find like ten other labels who have an amazing roster, and I've never heard of, and I'm like 'why would anybody care about [my label]?', and I do feel an element of competitivism and defeatism, at the same time. But then, you don't wanna compete, you wanna support each other, and there have been opportunities where we've collaborated with people, and those have been probably the most rewarding things that we've done. But I don't think you realize how common you are? Do you know what I mean?

P24: Yeah, I think when you notice how many other labels are on Facebook you're like, yeah [sighs] ...

P23: Damn, that niche is not as niche.

P24: There's a lot going on.

Several practitioners remarked on this sense that, online, supply of new DIY music greatly outstripped demand – something that felt less relevant in local, offline spaces. There are 'twenty million bands', remarked one, 'and not everyone is waiting at their computers for new music' (P21). Notions of 'convivial competition' can be undermined by these feelings of defeatism and irrelevance.

In outlining his concept of convivial competition in academia, David Harvie is clear to identify the marketization of higher education in the UK (and the resulting competition for students' fees) as the primary threat to this sense of researchers aiming towards shared goals. 'Within such an environment', he argues, 'it is not surprising if individual researchers and research teams co-operate less with rivals, and become more aggressive in claiming "ownership," i.e. enclosing, of ideas' (2004: 4). What makes competition 'convivial' is, in part, an assurance that the market is not intervening between competitors. One way to assure this might be to more explicitly engage in forms of collective ownership. With some notable exceptions, this has not been a very common strategy in DIY. In the next section, I continue to assess the strength of collectivism in DIY scenes, with a specific focus on the different kinds of ownership that are enabled by social media.

Taking ownership of DIY on social media platforms

The process of music-making is deeply social, even if it is not something that always takes place as a group activity. The creative choices made when writing and performing are selected from 'the particular universe of possibles' within a given field of musical production (i.e. a scene or genre) (Toynbee 2000: 40). Similar to Becker's 'art worlds' thesis (1982), this suggests that music-making is a kind of collective production, both at the point of origin (composition, performance etc.) and in the ways that music is received and used. Yet, in the music industries, emphasis is placed on the musical work being the 'intellectual property' of a single or a few individuals rather than of the field. Copyright law supports this understanding of creativity, with royalties affirming songwriters (and, for mechanical royalties, performers) as the owner-creator of the musical work and the ones who are therefore enabled to transfer those rights to record labels and music publishers.

Previous DIY scenes have occasionally attempted to find alternatives to this model of individual (or small group) ownership, recognizing its problematic consequences for participatory parity (i.e. the hierarchies discussed in the previous section). The final (to date) issue of seminal zine *Riot Grrrl* closes by highlighting that 'this name is not copyrighted ... so take the ball and run with it!' (in Darms, Lisa and Fateman 2013: 56), emphasizing collectivism by handing over the reins to whoever might want to take them up. Similarly, the DIY post-punk band Scritti Politti were initially reluctant to define where their band ended and their social circle began. In a fanzine interview circa 1979, their singer and founder Green Gartside outlined: 'The idea is that substantial decisions about what the group is doing are made by a larger number of people than actually pick up instruments at present' (quoted in Reynolds 2005: 182). In these ways and others, DIY practitioners have aimed to perform with a 'voice' that speaks on behalf of more than just the three or four members of a band.

But the majority of DIY practice has involved artist–audience relations that echo those found within rock and pop music more generally. The frequently espoused notion of the rock band as a 'democratic unit' within DIY (emphasized particularly by US hardcore and post-hardcore bands) has positive dimensions, but also serves to emphasize the binary nature of group membership (in or out). The status of bands and artists as *named* entities within the scene binds the musical work in a mode of ownership that inserts a claim of recognition in-between musical practice and the community. The distinction is subtle but important. The music doesn't belong to the scene; it belongs to a specific band or artist who belongs to the scene. Ultimately, these entities can leave the scene, taking their music within them, along with the collectively produced symbolic meaning contained within it. (This is essentially the process captured in the notion of 'selling out', as bands accept money in return for a crystallized set of values and concepts that arguably – and speaking ethically rather than legally – was not theirs to sell.)

As outlined at the top of this chapter, the internet's capacity to draw together contributions from users across the globe at high speed has brought about new ways of imposing order on creative practice. This section considers the means by which DIY practitioners assert ownership over their practice online, the extent to which this impacts on feelings of community (and the 'convivial competition' discussed above), and the consequence of these activities on DIY's capacity to offer valuable cultural resistance. I argue that for the most part contemporary DIY practitioners (especially bands) are inclined to act in ways that are compliant with social media's 'brand' logic and that new potentials for collectivism are under-utilized and relatively unexplored.

When considering their online practice, my interviewees would often reflect on their earliest experiences of the internet and the previous kinds of DIY (and non-DIY) music culture they had encountered there. For most of them, these early experiences pre-dated the centralization that characterizes the present-day internet. Whilst using MySpace was a formative experience for many, this was part of a larger ecosystem that included blog networks, genre- and scene-specific news sites, forums and message boards, and substantially more email usage. Forums and message boards were remembered especially fondly. In Leeds, there had been at least two active local music forums, which served as a hub for organizing and sharing: 'I think the forum at that point was a very, very fertile place for people to, like, bounce gigs off each other ... this understanding that there were so many people doing similar things, felt like you were part of something, I guess' (P22).

This memory of being 'part of something' suggests that forums are (or were) a clearly bounded space in which individual activity was contained within a collective effort. This sense of collective effort wasn't limited to local forums. For example, the indie-pop forum Bowlie (later renamed Anorak) served through the mid-to-late 2000s as a hub to connect activity across this international scene, with sub-forums for key UK cities (and other global regions) and for discussing specific aspects of indie-pop practice (promotion, songwriting, artwork etc.). These kinds of sub-divisions helped limit problems with spam and unwanted or irrelevant posts.

This online architecture of multiple, disparate websites is now largely gone, rendering many of these old sites either defunct or derelict: 'If you look at a post from about 2006, there'll be like pages and pages of discussion on the [Leeds DIY listings] forum, just in the community around that, and all the gigs and like, genuine enthusiasm that has migrated totally to Facebook now' (P12).

I don't mean to idealize these message-boards, which had hierarchies of their own. But they were free of direct corporate governance and moderation, and their independence marked a continuation of DIY practitioners' historical tendency to establish and operate within distribution channels under their own control (with Rough Trade et al.'s famous 1980s indie distribution 'Cartel' as arguably the apotheosis of this kind of approach). Where the old system of forums and fan sites operated as an online equivalent to these channels, with power divested amongst a collection of small sites owned and operated by practitioners, online DIY activity subsumed within Facebook constitutes the concession of a large degree of distributive control. In terms of the relationship between the community and the individual, this means that much individual online activity is no longer contained under the canopy of a DIY distribution channel.

The closest contemporary corollary to the forum, in terms of platform offerings, is the Facebook Group. Groups offer some of the 'boundedness' of forums and provide many of the same features (discussion threads, the ability to operate a selective membership policy) although without some of the subtleties (most notably they lack the capacity to create specialist sub-forums). However, these Groups do not provide the same home for DIY music sociality as forums once did: 'They're over-saturated with constant posts about gigs, or people, or bands, or spam. And people don't read them. I don't read those groups. I don't think a lot of people do. It's a really messy way of doing things' (P19).

When Groups came up in my conversations with practitioners, they were often immediately associated with 'spam'. But definitions of online spam tend to centre on the communication being unsolicited and being sent indiscriminately (Chandler and Munday 2016a) – quite different from Groups which practitioners have voluntarily joined and which constitute a discriminate audience with an express interest in the niche subject at hand. The sense of 'spamming' then results not from the content itself, but from an excess of self-promotion which accrues here and renders the Group ineffective for other social purposes.

Scrolling through the feed of the most prominent Leeds DIY Facebook Group in August 2016, I found that out of the twenty most recent posts, sixteen were drawing attention either to a new music release or to an upcoming show or event, alongside one request for a sound engineer, one 'band members wanted' post, one update on a new rehearsal studio construction project and one graduate researcher looking for archive materials (not me!). A small sample from the most prominent UK DIY Group featured the same ratio of self-promotion (sixteen posts out of twenty). The assumption that nobody reads Groups re-enforces the sense that self-promotion isn't particularly harmful, which in turn renders the Group less readable. The end result is a widespread belief that Groups are inherently of limited use: 'A lot of the people were on a [big] Group a couple of years ago that kind of died a death for some reason and I think since then people have been jaded about Groups and don't think that it works' (P1).

The predominance of self-interest on Groups is at least in part exacerbated by platform architecture. Because of the high proportion of promotional material shared on these Groups, many practitioners alter their notification settings from 'All' to 'Friends', meaning they only receive notifications from the Group when their existing friends post or comment in there. This means that existing social networks (and their associated hierarchies) are reinforced, even in groups that are intended to be recording a wider range of scene activity.

The primary means of managing the online presence of music projects is via Facebook Pages. Because Pages are a business-oriented platform, they tend to normalize an 'enterprise discourse' (Banks 2007), seemingly at odds with the ethics of DIY music. I cover that aspect in Chapter 6. But my focus here is on Pages' capacity for collectivism, and in particular in their relationship to Groups, or other similar messageboard-style options. One interviewee conceded that Pages are 'less chatty' than other options (i.e. Twitter) and provide fewer opportunities for sociability (P21). Why then do practitioners opt to make Pages for their DIY musical projects and thereby construct a more obviously competitive relationship between practitioners?

As well as the obvious communicative convenience, Pages offers a sense of validity and legitimacy – their ambivalent utilization by DIY practitioners is a continuation of DIY's ambivalent relationship to mainstream popular music practices. In Chapter 2, I suggested that DIY tends to emulate (because it finds communicative potential in) the 'core units' of popular music, such as the record, the band and the live show. Arguably the Facebook Page (and increasingly, the Instagram account, Spotify presence etc.) is a new 'core unit' of popular music – perhaps the first since the music video – and one which possesses attributes of permanence and uniformity that make it highly suitable for establishing one's self as a musical entity.

For DIY practitioners, setting up band accounts is the way that musical activity becomes 'real', even if it precedes any music-making taking place:

> Yeah, [band member] set us up a Twitter before we'd even had a practice, which feels really ill-advised, haha. I remember we hadn't even decided on the name, I wanted to call it something else, and then they set up a Twitter and I was like 'ok, well I guess that's what we're called now'. [...] I don't know why, I think possibly it was like a thing that meant we had to exist, because they'd ... I guess we were talking a lot about it, and then he set that up, and so it was kind of a driver to do stuff. (P21)

In this way, present-day DIY is in thrall to the norms of popular music-making in similar ways to its predecessors: legitimacy is achieved by 'enclosing' the activity within a band name. I will return to some broader consequences of this 'enclosure' in this chapter's conclusion. But I do think that, within the limitations of these frameworks, there are some kinds of existing DIY activities that demonstrate the political value of alternative uses of platforms. In the rest of this section I explore these efforts towards more convivial approaches.

The first kind of practice is that which explores the beneficial effects of anonymity. Anonymity online is today often associated with negative aspects of the internet (e.g. the notion of 'faceless' trolls acting individually or collectively to harass and harm others). This marks a change from an earlier optimism that anonymity and pseudonymity might allow for new forms of identity-formation (Turkle 1995, Kang 2000), and online anonymity continues to benefit vulnerable people who might have good reason to fear revealing their name and personal information. Like the internet itself, anonymity is inherently neither perniciously anti-social nor miraculously pro-social, even if some recent studies find a connection between anonymity and aggression online (Levmore 2010, Zimmerman and Ybarra 2016).

Practices emphasizing anonymity in the context of DIY seem to provide room to de-emphasize the self-branding norms of the attention economy – offering a space that allows opportunities for self-realization outside of the need to be 'relatable'. Anonymity involves consciously passing up on the *visibility* that is necessary to secure social status in a music scene (Scarborough 2017: 166), as well as opting out of the 'visibility game' that characterizes much online activity (Cotter 2019), thereby reducing the scale of the hierarchies discussed above. One female interviewee enjoyed using SoundCloud because it allowed them to put forward a minimal amount of profile information, leaving the listener with 'just waves' (P28) – remarking on the distinctive 'waveform' music player on the site. Another young practitioner shared a short anecdote about sharing their music online anonymously:

> I recorded an album on a Monday morning once, just to trick my friend with, and put that online. And that exists, and that still gets listened to. After uni I had nothing to do, so before bed I wrote five songs, and recorded them on a Monday morning, and was like 'I'll make a quick album cover and come up with a name', and then just put them online and go 'what do you think to this band' and he goes 'they're alright'. And that was the only reason I did it. I've not done anything with it since, they just exist online. (P14)

The activity described here may not seem particularly radical. But in the context of a growth-obsessed landscape, which emphasizes the accumulation of Likes and Follows for some unspecified future purpose, to have done something to 'just exist' as a standalone cultural text and then to have 'not done anything with it since' constitutes something unusual.

A second (and related) approach uses networks established on social media platforms in order to emphasize that DIY music constitutes a 'commons' rather than a site of competition, using features of platforms that are not based around notions of ownership. One practitioner was keen to emphasize the benefits of de-centralizing their activity in this way:

> It's not really a proper site that does most of the work for [our skill-sharing collective], it's just when people use #[collective name] and then you look that up [on Twitter] and you realize: oh my gosh, there's someone in Indonesia, and someone in Finland, and in Costa Rica, and they're all hashtagging this thing. […] It's become this thing where it's a skill share, and a source share too. Like we tell people how to work outside the male-dominated promoting system, and even how to write funding grants, or even just how to use your four-track [recorder]. There's always someone who's willing to talk to you about something, because there's so many women now. (P28)

Here the sense of scale that can make online competition so daunting and un-convivial is repurposed as a source of confidence, and an assurance that there's 'always someone who's willing to talk to you'. This practitioner also identified their practice as continuing a 'punk' lineage and also as oppositional to 'proprietary' forms of organizing:

It is really punk-y that way because people are like 'you should have a proper place' but I was like, I think by not defining boundaries of things, like 'this is the home of [collective]', and just saying it's a place you belong to by stating you belong to it, and that's it, it takes care of it. Because there's no proprietary thing about it, it's just something to embolden you, and something you can use whenever you want. And a lot of people were like 'what if someone uses it, or does something … ' and I'm like 'what if they do? so what?' […] I think having an account, even a [collective] email account is kind of not so great. I really wish that people would always write everything publicly, and then we could all help each other, instead of asking me for something, and then me refer. Cos I'm just an in-between person. (P28)

So, while neither anonymity nor a hashtag-based 'commons' offers a comprehensive solution to problems of DIY ownership online, they might offer imperfect means by which to bypass some of the most self-promotional aspects of social media. They hint at future directions for DIY, should the aim be to emphasize collectivism over individualism. (Possibilities for DIY's future relationship to social media platforms are considered more fully in Chapter 8.)

These practices are exceptions, though, to the general trend within DIY of adopting tools and practices as they become normalized within mainstream popular music. DIY's interest in an ambivalent emulation of popular music means it often follows paths forged by star artists, albeit often with heels dragging, in search of communicative pertinence.

Social media and social authorship

In this chapter, I have suggested that the use of proprietary tools such as Facebook Pages, which focuses attention on reputation and reward, exacerbates the more individualistic tendencies in DIY music practice and diminishes the sense of community within scenes. In this short concluding section I want to highlight social media's relationship to another obstacle to collectivism, which is not exclusive to DIY: namely, a thorough and widespread disavowal of the truly social nature of musical creativity.

As noted above, Jason Toynbee's work on music and copyright makes a strong case that musical creativity is best considered as 'social authorship' (2001). This means that, while making music does contain some degree of 'originality', this is of a very limited kind and primarily involves 'recombining symbolic materials from a historically deposited common stock'. We all make music *together* even if the actual assembly of those materials is often done alone or in small groups. But, as he continues: 'If the practice of authorship is social and not romantic, a significant issue still remains: the persistence of a strong form of romantic discourse. In popular music, [...] the common sense of musicians, fans and critics is that the originality and genius of the music maker is paramount' (2001: 3). Toynbee notes that this discourse primarily works in favour of large cultural industries corporations.[1]

♪ 'To be objectified' – Jeffrey Lewis

Platforms like Facebook, and particularly Facebook Pages, do tend to reinforce romantic, individualistic notions of music-making. I have noted that this is fairly compatible with DIY's historical tendency to emulate popular music forms and DIY practitioners' continued desire for recognition at the artist or band level. But platforms also extend this individualization further, and this is potentially deleterious for DIY ethics, since the activity 'captured' and individualized by platforms is not just akin to the 'old' musical commodities (i.e. songs, recordings), but now extends a bulk of the 'everyday', extra-musical activity of the scene. All scenic activity, when re-presented online, is inherently positional, since it is always shared from someone's account; it always belongs to the uploader of the content, rather than to the scene in general. This does not inherently lead to hierarchy or to excessive self-interest. But it does diminish any sense that a scene is itself a work of 'social authorship', rather than a series of distinct individual contributions for which credit can be assigned. It posits the scene as network, in which there is no collective 'middle' that sits between practitioners, only the mappable connections between individual actors.

Platforms' emphasis on the individual user is partly economically motivated – since they are the primary object around which surveillance and advertising are organized. But the notion of individual 'nodes' is also central to the very idea of networks; even groups and corporations must be represented by a single account (or by a series of single accounts). The fluidity, flexibility and inherent sociality of everyday communication are distilled into rules of accountability and assignability – instrumentalized knowledge about who did what, when and where. A key threat that social media poses to DIY music is that these economic and technological

---

[1] This is primarily because 'romantic discourses of authorship [...] have legitimated the most important commodity form in the symbol world – copyright' (Toynbee 2001: 3). Although copyright has not been central to my analysis thus far, the notion that music can be assigned to a particular author-owner of course underpins the possibility of it being sold as a commodity.

principles are fundamentally at odds with real experiences of togetherness. Internal relationships in DIY scenes rely on a conviviality and a generosity of spirit which platforms currently struggle to meaningfully accommodate.

Of course, DIY scenes do not only communicate internally. And part of what commodification offers is the capacity for cultural texts to travel far and fast, to unexpected and unfamiliar places. In the next chapters, the politics of DIY and social media are considered with regard to these wider relations.

5

The public

Performing politics and elucidating difference

The previous two chapters have examined how DIY music practitioners use social media in terms of their personal identity work, and at the level of their inter-scene communications. In this chapter we continue moving 'outwards', in order to look at how DIY practitioners use social media to engage with people outside of the scene – that is, the general public. The notion of a single 'public' can be an unhelpfully totalizing abstraction and politically detrimental insofar as it conflates the competing and contradictory goals of what are in fact multiple 'publics' (Fraser 1990). But I use the notion of 'the public' here because it does meaningfully reflect how DIY music practitioners often informally differentiate between inside and outside of their scene. Practitioners' perspectives on the world 'out there' – which, of course, they mostly live and work in – play an important part in defining the scene, and the kind of boundaries and borders that are established around it.

I'm treating this as something slightly different to the idea of popularity, which is explored in the chapter after this one. Popularity, in my critical framing of its role in DIY, relates to artistic tensions between expressing one's self and representing others, to problems of how to gauge success and failure, and to DIY's real and imagined engagements with the popular music industries. The concerns considered in this chapter are related, but are in general less grandiose, and operate at a level closer to the 'everyday'. Bands often don't get much of a choice regarding whether they are popular, but even unpopular, virtually unknown bands will necessarily make some choices regarding what rooms to play, and will have some preferences as to who should and shouldn't be in that room. They have to deal with unavoidable practicalities of where to be and who to be with, and their choices are not limitless. One of the main 'publics' that DIY practitioners encounter, then, are other local music scenes and their local music infrastructures – the live venues, publications, rehearsal spaces and so on, that are available to them. Interactions with other scenes and other audiences, and particularly their online manifestations, are a key concern of this chapter.

In making decisions regarding how, where and whether to engage with a 'public' outside of their scene, practitioners negotiate a longstanding tension in DIY music between insularity and openness, both of which can be ascribed value in terms of cultural resistance. On the one hand, some degree of insularity might be required in order to incubate fragile, developing cultural forms, or to protect vulnerable or marginalized groups in the scene. An excessively insular scene, however, might be seen having a limited capacity for cultural resistance, as well as potentially tending towards elitism, or plain mundanity.

The riot grrrl scene of the early 1990s offers an exemplary manifestation of this tension and its relationship to cultural resistance. Some riot grrrl artists would utilize misogynist slurs in their work and would sometimes paint them onto their body, as a means of reclaiming ownership over the people and bodies to which those words referred. Such a strategy requires that the audience has some shared understanding of this process of reclamation; Dave Laing notes that the danger with subversions of erotic performance in punk music is that they might 'simply [...] be read by the omnivorous male gaze as the "real thing"' (Laing 1985: 117). DIY cultural expression runs the risk of being misread by audiences that are on a different wavelength in terms of musical and cultural expectations. Accordingly, some part of riot grrrl's 'girl revolution' took place outside of 'the public world, the world of men' (White 1992), in a context where practitioners could support each other in self-actualization without interference. Zine distribution networks helped to support this grrrl-to-grrrl communication, and flyers were used to help filter the audience of live shows (e.g. by declaring 'girls to the front') in order to create a space where the presence of outsiders was clearly delimited.

This kind of insularity made certain types of resistance possible, but may have set a high threshold of 'collegiate erudition' (Gottlieb and Wald 1994: 271) that excluded many would-be riot grrrls from participating. And, for some practitioners, powerful cultural resistance resided in those moments of friction in which they came face-to-face with those who were either opposed to or nonplussed by them. Liz Naylor, who booked and managed the 1993 Bikini Kill and Huggy Bear UK tour, argued that these clashes constituted the place where ideologies might meet and interact. She recalls an audience of 'pissed up lads' – in her terms, 'the wrong audience' – creating 'sort of a riot' at one show, and reflects: 'I kind of liked those moments in a way because I think they're more challenging [...] I quite liked some of that confrontation' (quoted in Downes 2009: 176). Such events can be seen as 'evidence of the authentic challenge riot grrrl represented to the social order in enabling punk-feminism to confront the "wrong" audiences and places' (Downes 2012: 230).

So, DIY music scenes have long been concerned with notions of right and wrong audiences, and the times and places at which it might be either fruitful or detrimental to attempt engagement with other scenes and other publics. This tension between insularity and openness does not lend itself to a permanent resolution; it is a consequence of DIY's ambivalent relationship to popular music (see Chapter 2). DIY music history is characterized by an oscillation between seeking insularity and seeking openness, and different scenes have taken approaches that reflect their distinct social positions, cultural ambitions and their theorizations of political change.

A critical history of online communication might similarly be told in terms of such an oscillation. In the 1990s and 2000s, the internet was frequently framed as a technology that would engender new possibilities for deliberation and democratization, and thus for rejuvenating the public sphere (see Papacharissi 2002). More critical assessments, of course, have been plentiful, and many of them have problematized the internet as a communicative space that is either excessively insular or excessively open.

Perhaps the earliest and longest-standing concern relating to excessive insularity in online communication has been the question of who does and does not have access – later framed in terms of the 'digital divide' (Jenkins 2008, Jenkins et al. 2006). For Michael Hauben, a young scholar and internet user writing in 1995, to witness the deliberative online communication taking place on Usenet discussion groups, was 'to see the democratic ideas of some great political thinkers beginning to be practiced'; a substantial mitigating factor was that this new deliberative realm was 'only open to those who either have it provided to them by a university or company that they are affiliated with, or who pay for it' (Hauben 1997: 319). In the early 2000s, as social media platforms began to prioritize connecting users to their existing offline social circles, scholars and commentators became increasingly concerned that the internet was being divided into 'echo chambers', which might have negative consequences for deliberative democracy (van Alstyne and Brynjolfsson 2005, Sunstein 2001, 2004). The related notion of the 'filter bubble' (Pariser 2011) describes a similar consequence, but with platforms themselves playing a more causal role. Here it is algorithms rather than the social groups themselves which are understood to be the motors of segmentation. Recent fears around insularity online have also been connected to fears over political and cultural 'radicalization', whether as a consequence of YouTube algorithms leading viewers to extremist content or as a result of excessive exposure to niche communities on messageboards like Reddit and 4chan (Robertson 2019).

Concerns around excessive openness are perhaps even more longstanding. Fears that the popularization of the internet (and its predecessors) would destroy the specific social conventions, or 'netiquette', that had enabled it to flourish, are almost as old as the internet itself. As early as 1985, Usenet 'old-timers' (who had, at most, five years' experience on the site) felt that the anarchic and offensive behaviour of a 'new wave of hot-tempered users' was betraying the ethics and spirit of the space (Lueg and Fisher 2012: 28). The inverse of the 'echo chamber' critique is found in accounts that emphasize just how easy it can be to cross social divides and cultural contexts online, and the ways that this capacity can be harmfully exploited, both discursively (e.g. trolling) and technologically (e.g. through DDoS attacks). The 'scalability' of social media means that private or personal incidents can quickly become immensely public (boyd 2011), and this can lead to uncivility and abuse, which often follows existing patterns of gender and racial discrimination.[1] For marginalized people especially, online platforms do not always – or ever – feel like an 'echo chamber' (Gray 2012, McMillan Cottom 2015). This kind of openness creates immense challenges for platforms (Gillespie 2018, Roberts 2019), whose moderation policies have been criticized from a variety of political perspectives.

As with DIY's relationship to insularity and openness, the tension outlined above is not one that requires or permits solving – it is an inherent contradiction that shapes ongoing engagements between users and platforms. The variety of capacities and features offered by contemporary platforms is sufficiently large that social groups seeking specific communicative 'shapes' and 'sizes' will tend to find the tools that do the job. Accordingly, there is no inherent best communicative architecture which conceptually precedes the arrival of actual users – with all their social, cultural, and economic baggage – into that space. This does not preclude the normative assessment of social media platform design. But such assessment ought to fully account for the fact that platforms are reflexively and continuously constructed in tandem with users who will bring different reasons to value insularity or openness in their communication.

In the case of contemporary DIY music, usage of social media strongly tends towards valuing and seeking insularity. In the next section, I show that this reflects a broader allegiance to discourses of safety, and the important role of 'safe(r) spaces' in DIY scenes. In the third section, I consider how these aspirations are further enabled and sometimes exacerbated by social media. I also discuss the negative consequences of an excessively insular scene, primarily in terms of diminished opportunities for transcendent and transgressive musical experiences. As a consequence of this insularity, full-scale online arguments with other music scenes weren't very common, but I close the chapter by presenting a case study of one of those arguments. This case study highlights that DIY's political reach is substantially impinged by practitioners' low faith in the possibility of social media platforms providing a space for meaningful deliberation with others.

---

[1] Zoombombing – the intentional disruption of live video events – seems to be the latest iteration of this tendency. As the COVID-19 pandemic moved much live music activity online in early 2020, at least one UK DIY show was interrupted (and ultimately halted) by racial abuse towards performers in this way.

Creating safe(r) spaces

The concept of a 'safe space' (or, sometimes, 'safer space') was fundamental to Leeds DIY practitioners' understanding of their scene. The term was employed by a majority of my interviewees, who used it to demarcate something specific about the DIY scene that differentiated it from both other music scenes, and from the wider world. The concept has its roots in feminist activism of the late 1960s, where the attendance of some 'consciousness raising' meetings was limited to only women.[2] The logic of the safe space was based on an understanding that many social and cultural norms were rooted in sexism, and therefore a space which allowed for the temporary escape of those norms was beneficial in developing individual and collective political awareness. The notion of a safe space moved from feminist activism to the gay communities of New York and Los Angeles in the 1960s and 1970s, where gay neighbourhoods would be policed by activists who would meet homophobic abuse and harassment with physical force (Kenney 2001). Whereas early feminist safe spaces were intended to discuss and bring forward new solutions in terms of political action, in the gay community the safe space *was* the solution, ensuring that gay identities could be publicly performed within that space. Even though they are often considered 'safe havens' rather than 'sites of resistance' (Myslik 1996), safe spaces have assisted groups, movements and individuals to develop and practise autonomy, to think and act critically, and to temporarily forego oppressive and discriminatory cultural norms.

The most substantial attacks on the concept of safe spaces in recent years have come from the traditional press (mainly right-wing broadsheets) in the UK and United States (Gosden 2016, Travers 2017), and also – somewhat ironically – from the growing presence of identitarian politics across much of the globe. In this negative coverage, safe spaces tend to be misrepresented in two main ways. Firstly, they are presented as an encroaching threat to free speech, foisted unwillingly on the silent majority by a vocal cabal of 'social justice warriors'. In actuality, safe spaces usually have distinct borders either temporally or physically (e.g. a single room is designated as a safe space for the duration of a meeting), and almost always require the participation and consent of those present. Secondly, these critiques equate safety (and the accompanying concept of 'trigger warnings') with an unwillingness to engage and debate with different viewpoints. This critique is often closely linked to the assessment of millennials as sheltered and easily upset (Hosie

---

[2]The question of who poses a 'danger' to who is a subject of fierce, ongoing debate. The construction of these safe spaces has, in some radical feminist movements, involved making problematic, essentialist distinctions between trans women and women assigned female at birth (Hines 2019, Withers 2010).

2017). In fact, safe spaces may well have the potential to bring about *greater* engagement with difficult subject matter, especially in educational settings (Mayo 2010), creating an environment in which fear and trauma can be replaced by curiosity and critical exploration (Thompson 2017). Thompson argues that, at their best,

> safer spaces practices *make life difficult*: they require us to attend to often unarticulated power dynamics and hierarchies that exist 'in here' as well as 'out there.' They require us to become sensitized to forms of encounter that we are too often desensitized: to soften to that which we are otherwise hardened. They force us to rethink common-held understandings of violence and harm; and to take seriously the action of speech-acts.
>
> (2017)

Potentially, then, the notion of 'safety' is rather misleading. The concept might be better understood as a means of prioritizing, in social theorist Nancy Fraser's terms, the 'recognition' of marginalized identities over the comfort of non-marginalized people who are the primary beneficiaries of the kinds of 'common-held understandings' identified by Thompson.

Many contemporary DIY music venues have a safe(r) space policy, including the two most prominent cooperative DIY venues in Leeds (Wharf Chambers and Chunk). Practitioners pointed to this policy as something that differentiated these spaces from other (non-DIY) venues, and also from the other pubs and clubs that shared Wharf Chambers' city-centre locale. The existence of venues such as Wharf and Chunk provided clear 'homes' for DIY music in Leeds and offers a marked contrast from the experiences of a previous generation of DIY practitioners in the city. Ros Allen, bassist for two seminal bands in the late 1970s post-punk scene (The Mekons and Delta 5), recalls regular conflict with the far-right fascist organization National Front, resulting in 'a lot of violence and aggression' and, on one occasion, 'a gang trying to disrupt a Mekons gig by goosestepping toward the front of the stage, clearing the dance floor and Seig Heiling at the front of the stage' (Allen 1996). Such events may well have taken place in The Fenton, a pub with post-punk pedigree that still operates as a live music venue today, but which has fallen from favour in the contemporary DIY scene. Along with other similar spaces, sometimes characterized as 'room-above-a-pub' venues in the city, it was seen as a reserve option and would be hired by promoters only if Wharf Chambers and Chunk were unavailable.

♪ 'Never been in a riot' – **The Mekons**

Online tools have generally been used to increase the sense of safety, stability and insularity in the DIY scene. For one thing, they allow performers and audience members to conduct informal 'research' on venues and bands. One practitioner recalled their earlier DIY experiences with 'entire tours booked on the phone', in which 'you'd just end up in some really crazy situations', and contrasted these with present-day tours where 'you can see the venue [online] and get a feel for it' before arriving, or even before confirming the show (P13). This is achieved in part through functionalities that aren't necessarily music-oriented, such as viewing photos of venue interiors on Google Maps, as well as photos of their exterior locale on Google's 'Street View', and checking venue ratings and reviews on Facebook, Google, Yelp and Tripadvisor. These social media functionalities seek to replace exploration with information, in order to avoid unwanted surprises.

The internet has also substantially altered the process of promoting small DIY shows, in ways that tend towards insularity. Although promoting gigs with physical posters was considered 'more rewarding' than digital promotion (P19), for various reasons – fewer shops allowing posters, a perception of declining efficacy, plus increasing costs relative to online promotion – DIY promoters were increasingly reliant on Facebook Events as the primary tool for announcing and publicizing shows, as well as updating potential attendees with information such as ticket price and stage times. This reliance was so all-encompassing that my interview participants were often unable to identify how people outside the scene might possibly find out that there was a show on. Other online methods, and particularly local music forums (addressed in Chapter 4), have fallen by the wayside. Facebook Events were reluctantly acknowledged as the best tool for the job.

♪ 'You are invited' – The Dismemberment Plan

This is despite the fact that the visibility of Facebook Events tends to be highly limited to those within the friendship group (and the 'Friends of Friends' network) of those playing and putting on the show. Indeed, although they are public events, their visibility can be quickly diminished if one suddenly finds one's self outside of the tight circle:

> On social media particularly, one big problem I had is that I'd blocked the person who I'd been seeing and broken up with, and also blocked their two housemates, so that I didn't see anything being said to my ex which would be upsetting to me. But then suddenly because I'd blocked all these key players in the queer DIY scene, loads of events weren't even

visible to me, and people would be like 'oh, are you going to this thing on the 25th? It's on Facebook'. And I'd search and nothing would come up, and after a bit, I'd think 'ah, it's probably that the Facebook event has been created by one of these people that I've now blocked'. So that was a bit difficult. [...] And so I feel like nowadays, to effectively participate in the DIY scene to its full potential, you have to be on social media, because that's how stuff is advertised. (P9)

The emphasis on constructing and maintaining a safe space was the dominant discursive framing of DIY's relationship to the wider public. In this regard, it was undeniable that the scene was in a much stronger position than ten or fifteen years ago, in terms of having dedicated spaces to call its own.[3] And this, in part, informed an insular approach to promoting shows online. As I have outlined, however, a tension between insularity and openness is fundamental to DIY's relationship with popular music. Many practitioners had concerns about the consequences of a scene that was excessively inward-looking. In the next section I unpack some of these concerns and focus on the impact of insularity on the political-aesthetic capacities of DIY music.

The benefits of friction?

Many DIY music practitioners in Leeds were aware that an emphasis on safe spaces might have a harmful effect in terms of inclusivity measured across some demographic axes, such as class and educational status:

> That [safe space policy] does give you a degree of protection, but then it replicates a problem of 'we're only inclusive to people who are already in.' And it is brilliant, it's brilliant and a lovely, amazing safe space where we're all really on it, and we all protect each other, but you're only allowed in if you already know how to play by these rules, and you already know some people in it. (P9)

---

[3]The stability of Wharf Chambers is not necessarily a common feature of local DIY scenes in the UK, although newly established venues such as DIY Space For London (DSFL) are increasingly following Wharf's lead in providing a single, stable space for DIY shows and events within a city. Additionally, Wharf Chambers' longer-term future is by no means assured, as a leased property within a highly desirable area. The increase in construction of city-centre housing means that live music venues are increasingly threatened with closure due to noise restrictions (Davyd and Whitrick 2015).

In this regard, though, most people seemed to think that the current trade-off was worth it. Many practitioners felt, understandably, tired of acting as the educator to people who they felt it was not their job to educate. Practitioners are not just (unpaid) stewards of the scene, but beneficiaries of it, and the never-ending work of being inclusive to ignorant newcomers might affect their ability to fully participate:

> [DIY] is kind of an outreach thing, and making it accessible to other people who might benefit from it. But also, if it's the way that I make friends and the way that I meet partners, I kind of just want it to be a nice space for me to have fun, where I can just go out and not worry about people being dicks, and not have to make too much of an effort with people I don't know, and have a safe space just for my own enjoyment. (P9)

However, some practitioners reflected on the potential downsides of a kind of safety which, whilst not a direct consequence of the safe space policy, was strongly associated with the stability of Wharf Chambers in particular. They used the notion of 'friction' to outline this:

> It's a hard thing to talk about because I'm in a very privileged position, but I know people who the safe space policy is there to protect, much more than me, cos in the street they feel threatened 'cos of their gender or the way they look, and those people often want friction [...] in a gig, as well, which doesn't exist as much in a space like that. [...] Don't get me wrong, this is me really nitpicking cos like, I think I couldn't really wish for a better space than Wharf Chambers for what I'm interested in, so this is complaining about something ridiculous in a way, but there's something I miss about seeing a very extreme band, in the kind of space that doesn't make sense. (P12)

This was a common theme within interviews: an identification of insularity as having very significant benefits to the scene and also having some subtler, less significant downsides.

In this section I examine those downsides, which I understand to be a consequence of the inherent tension between insularity and openness that I outlined at the start of this chapter. I consider what kinds of value might be found in 'friction' between DIY and other scenes, whilst retaining an awareness of the discriminatory social norms that necessitate constructing safe spaces in the first place.

Such a conception of generative friction would, for DIY practitioners, not include violence and intimidation of the kind experienced in Leeds by post-punk musicians of the 1970s and 1980s. But it might involve feeling scared or threatened (not as a result of discrimination) in the context of musical performance, and it might mean taking steps to understand different musical cultures as aesthetically different but ultimately comprehensible, a goal that could potentially be achieved through DIY offering a more diverse line-up in terms of musical styles. One practitioner warned of the negative effects of an environment in which 'everyone in the room has already agreed before they get there that this is how things should be done'. 'Yeah, it's peaceful and everyone feels comfortable', they acknowledge, 'but where's the evolution of the idea?' (P13). Such a rhetorical question implies that 'evolution' (whether political or aesthetic) requires *friction* of some kind – a process of synthesis resulting from the existent meeting the new.

The benefits of this kind of friction might also be gained by playing shows at other, non-scene venues. This was a viewpoint primarily held by the older, male practitioners (who are arguably less in need of the 'safety' offered by venues like Wharf) rather than the younger ones, and they drew on past experience to relate what might be lost in the transition to a safer, more insular scene:

> When you make music that you think to yourself is very poppy, playing in a weird pub [as opposed to Wharf Chambers] often highlights how different it is from what actually mainstream music is. I often was in bands that I thought were quite palatable to people in general, 'cos they had melodies and were fun sounding or whatever, in broad terms. And if you play somewhere like that, you realize that people are like, 'what the fuck is this?' Even indie pop, I think, is not that palatable to your average pub punter. It reminds you how you can quickly go down an alley, that the way you think about music changes, so that you see something very extreme and be like, 'oh, it's boring'. Being confronted by mainstream culture is quite important to figure out where you are in the world. (P12)

For these older, straighter, whiter, maler practitioners, this radicality of aesthetic might be especially important for identity construction. They cannot depend on the under-representation of their politicized identity as a means of imbuing their musical activity with alterity. However, I think there is a more generous reading of this argument, as highlighting the potential for excessive insularity to downplay the extent to which DIY music can be valuably critical of mainstream popular culture. A performance that creates friction might help practitioners consider the functions of their music as art (in the sense of a critical reflection of society), as opposed to its function as entertainment (as a pleasurable and enjoyable communication). To present one's music to an unfamiliar crowd, and to see one's music fail to engage them and nonetheless still find value in it, could allow for reflection on one's own subjectivity and the different 'psychic drives' that Georgina Born identifies might be at work within those who seek popular acclaim and those who value alternativity (1993: 237).

In contrast, younger DIY practitioners were far less interested in creating 'weird' performances or exploring the possibilities of being in the 'wrong' space:

> We got offered this gig recently from a very generic rock promoter [...] at this mixed, local band venue, upstairs in a pub. And it's like 'do we want to play that'? We were all joking like, we're too famous to play this now, and obviously we're not, but we've got a scene we want to contribute to, and it'd be like playing a gig to a group of strangers that probably wouldn't get it ... or at least, I'm not very good at playing guitar – and on the scene that's fine, it's kind of my schtick. And it's like, do we play it? (P14)

Being 'famous' is used here in a roundabout sense to denote a context in which communication is more likely to be successful; the image is that of the 'famous' rock star in simpatico with their audience, mirroring their fans' emotions and acting as their on-stage representative. The audience knows the songs, they know what to expect, they understand what is being aimed for and no additional framing work is needed: they just *get it*. In contrast to the communicative distance between them and the imagined 'group of strangers' at the pub, the same practitioner describes playing at Wharf Chambers as a place where communication is easy: 'The deck is stacked in your favour when you're playing at Wharf, it's a cool space with cool people, everyone's gonna be really nice, you'll have a great time, and your music's meeting the right audience, 'cos Wharf *is* your audience' (P14).

This practitioner is liberal in their usage of unqualified, positive adjectives to describe the experience of playing Wharf – it's 'cool' (twice), 'nice', 'great' and *'right'*. Yet there is a sense of nagging dissatisfaction (which is, again, accompanying a stronger, primary sense of gratitude to and support for this welcoming environment). The future tense creates a tone of assured prediction ('gonna'; 'you'll'), suggesting that there is minimal scope for unexpected events, either good or bad. In this presentation, Wharf Chambers is a near-frictionless space, mediating flawlessly between artist and audience, but lacking any element of the unpredictable.

Keith Kahn-Harris, writing on extreme metal scenes, argues that central to such a scene's construction is a tension between the 'mundane' and the 'transgressive' (2004). Transgression is the sense of 'exceeding' the body, and the sense of scenic activity as containing radical or 'deviant' potential; Kahn-Harris suggests these experiences are more common when first entering a scene. Mundanity is in some ways the opposite of transgression, but is required to prevent 'over-transgression', and to provide the structure that prevents the scene from disintegrating entirely. 'The problem', he argues, 'is that the experience of mundanity threatens to dominate the experience of the scene to the exclusion of all else' (2004: 112), as people seek to make the scene feel 'everyday'. In the contemporary DIY scene this 'everyday'-ness provides important feelings of safety, but for some practitioners it also brings feelings of mundanity.

Opportunities for experiencing unexpected, transgressive performances are rare in the DIY scene I studied, and are made rarer as a consequence of online practices. It is common practice to 'sample' a band online (through Bandcamp, say) to see if they might be to your tastes, before deciding whether to attend their show. One consequence is that people can quickly rule out attending shows outside of their comfort zone. Additionally, recordings might give an incorrect representation of the live experience, as one participant argued with regards to a band he enjoyed:

> So, if I was not particularly into experimental music and thought 'okay, I'll see what their Bandcamp sounds like, and it's this screechy lo-fi recording ... you wouldn't be able to discern what was going on particularly. That, compared to the experience of seeing them live, which is pretty mind-blowing experience cos they are so involved in what they do, and it's so loud and all the sounds are particularly extreme in different ways [...] I think that could actually make people think 'oh this is amazing'. [...] You can really limit yourself based on what you think something's gonna be like, based on the evidence that the internet provides you about this stuff. (P12)

An insular DIY scene is important because it provides opportunities for self-recognition, particularly for people who are marginalized. But when a scene is particularly insular, it can become mundane. Insularity also leads to a decrease in the number of situations where music might have a transformative power. This is true both within the scene, through a lack of opportunities for performative 'play', and beyond it, due to an unwillingness to risk playing to unfamiliar audiences.

♪ 'Nothing to say' – The Kinks

Mark Mattern's book *Acting in Concert* is a study of popular music's relationship to 'community-based political action', in which he proposes three main categories of political musical engagement: confrontational, deliberative and pragmatic. These categories of political action can manifest in either inter- or intra-group forms: in the case of deliberation, either group members 'use musical practices to debate their identity and commitments' or 'members of different communities negotiate mutual relations' (1994: 25). My research suggests that contemporary DIY music is highly focused on *intra-group deliberation*.

This isn't to say that DIY practitioners aren't looking outwards or that they have no interest in broader politics. But their musical activity isn't at the heart of that kind of outward-looking political participation. The sense that conflict and deliberation might be required to launch an 'authentic challenge' to dominant norms, as was felt by some in the 1990s riot grrrl scene, is far less observable in the contemporary DIY scene. Most people don't want to play to new and difficult audiences, and are happy to understand themselves as 'contributing' to the scene rather than branching out beyond it. It would be a stretch to suggest that an increase in DIY insularity has been caused by social media, and practices of online insularity do not ever fully 'secure' a scene from unwanted external interference. Indeed, platforms can enable such interference: Liz Pelly, in her assessment of social media's impact on US-based DIY, refers to police officers shutting down DIY shows in Boston 'brag[ging] about the fake Facebook accounts they used to find addresses for gigs' (2020). But social media's 'echo chamber' effect does work in tandem with this insularity – promotional techniques are more inward-looking (fewer posters, more Facebook Events), and platforms and services are often used in ways that increase predictability offline.

Overall, though, this focus on inter-scene deliberation means that communication with other scenes online is rare – especially with scenes that are not in some sense social or musical 'neighbours'. But I conclude this chapter with a case study of an occasion when this did occur, documenting two Leeds DIY bands having a fractious encounter with another band from outside of the scene, which illustrates the problems of social media as a setting for inter-scene interactions. Along with piecing together the narrative of this dispute through the online evidence it generated, I also conducted an interview with one of the DIY bands in order to discuss it in detail.

Case study: Arguing with other scenes online

In June 2015, the three members of a young 'queer punk' band from Leeds, all aged between eighteen and twenty-one, were organizing a short UK tour with another young Leeds band. Through a family connection with a band already on the bill, they were able to get both bands added to the line-up of a show in the West Midlands (around two hours' drive away). They were aware that the gig would be something different to their experiences to date playing in DIY venues, and primarily in Wharf Chambers: 'I think we're all coming from a DIY sort of scene, it was clear that this promoter wasn't really that kind of promoter. It was just a more sort of "clubby" promoter. But we thought, "yeah, just go with it anyway"' (P5).

Having confirmed the gig, the promoter then added another band to the line-up, whose name contained a strong, bluntly worded denunciation of Roma and traveller communities, including use of a derogatory epithet. The two Leeds-based bands were made aware of this addition through the title of a Facebook Event page being updated to include the new band's name. I refer to this new band as the 'pub rock' band, following my interview participants' description of them, although their musical and visual aesthetic placed them towards the heavier end of classic rock.

One of the Leeds DIY bands then privately messaged the promoter of the show on Facebook and asked them to remove the pub rock band from the line-up. The promoter acknowledged that he had himself found the name problematic initially but didn't remove them from the line-up, instead trying to placate both parties by vouching for the pub rock band's good nature. When the two Leeds bands made it clear that their attendance at the gig was dependent on the pub rock band being removed, the promoter opted to side with the pub rock band as they were local to the West Midlands and had an established following. According to one Leeds DIY practitioner, he was 'was more bothered about getting people in' (P3) than adhering to any moral principles.

Somehow, possibly via the promoter, the pub rock band got word that their name had upset one of the other bands on the bill. This caused a flurry of online communication, indirectly and directly, publicly and privately, between the various parties. All three bands concerned posted updates from their Facebook Pages. The pub rock band's post took a jovial, boisterous tone, expressing confusion and annoyance that another band had been offended; the two Leeds bands posted more circumspect statements explaining why they would not be playing the show. Following this, the social groups around the pub rock band and the two DIY bands (i.e. their friends and members of their respective scenes) were brought into contact with other, through links sent privately and publicly by the bands (and also possibly by doing their own searching), with wide-ranging arguments taking place on all three bands' Facebook Pages, and privately between bands on Facebook Messenger.

My interviewees considered this argument to have been, for the most part, unsuccessful and unhelpful:

> P3: I feel like there was one person I was arguing with who was more level headed than the rest of them. But generally, they didn't wanna, and we didn't wanna, you know ... we were both firmly where we were. And it just kept going, the same argument, for ages.
> P5: It was very circular. But I think when someone used the term 'the PC brigade', I sort of switched off.

They were, even before these events, very sceptical about the capacity for this kind of online debate to be productive. One participant 'switching off' at the phrase 'PC brigade' suggests that these are set positions within much-rehearsed arguments. This switching off reflects an assumption that the pub rock band (and their social group) lack self-awareness, which makes debate difficult and perhaps pointless. One participant suggested that 'I think they genuinely didn't think they'd done anything wrong' (P5). Another admitted that the pub rock band probably 'weren't like, terribly dangerous people or anything' but that 'when there's that hint of like sly racism that they don't fully understand, it makes you wonder what other issues they're a bit ignorant on and they're probably not fun to be around' (P4).

The Leeds DIY bands' error, according to the pub rockers, was to have taken offence to something that was not intended as offensive. The additional context provided to demonstrate this was, firstly, that their name was a reference to a movie, rather than a phrase of their own creation and, secondly, that the band were not themselves racist and did not have the negative perception of the Roma or traveller community that the band name would suggest. The pub rock band's social group wanted the DIY group to acknowledge this fluidity of meaning, but the Leeds DIY practitioners were unconvinced: 'Yeah, I didn't get that argument. Like, "we're nice people so we can't be racists," but then me personally, I judge someone's being racist on racist actions like choosing to call your band something racist, which is pretty straightforward to me' (P5). Whilst the DIY practitioners framed the conflict as ignorance versus enlightenment, the pub rock band framed it as common-sense pragmatism versus inauthentic moral hysteria.

The initial suspicions carried by the DIY bands concerning the moral character of their opponents were, they argued, soon justified by the use of homophobic language by a member of the pub rock band's social group, generating more bad feeling between the parties. The arguments on all three Facebook Pages petered out within a couple of days. Some weeks later the original gig went ahead, with the pub rock band playing and without the Leeds DIY bands. The two Leeds bands were able to find a replacement show in broadly the same area, promoted by a young, non-musician friend who was inspired by the events to organize and promote their first ever gig. The situation which necessitated its existence clearly imbued the show with an added political impetus.

The architecture of social media (specifically Facebook), as utilized by these social groups, impacted the initial friction and shaped the subsequent interactions in significant ways. It was the updating of the Event page by the promoter which initially informed the DIY bands of the line-up change, meaning that the plain fact of the offensive band name came first, in an automated notification sent to event attendees, without any additional context. When the young organizer of the replacement gig felt there was a threat of violence (after a member of the pub rock band told one of my interviewees that he would 'see you there'), they were able to remove the address from the Facebook event, and instead informed potential attendees privately about the gig's location. Most significantly though, the bands were able to use Facebook to bring their respective social groups into contact with one another, crossing cultural, social and geographical distance in order to bring two disparate groups together.

Much has been written on social media's capacity to facilitate 'bridging' (i.e. linking social groups) or 'bonding' (strengthening existing social ties) – using terms employed by political scientist Robert Putnam (2000) to distinguish different ways in which social capital is developed. Social media is particularly effective at bonding (Smith and Giraud-Carrier 2010), but also provides the technical apparatus and the social context whereby latent ties (i.e. potential acquaintances) can become actual connections (Haythornthwaite 2002, 2005). The online interactions between the social groups described here are certainly closer to bridging than bonding, inasmuch they create weak, temporary ties which did not previously exist. However, the bands' usage of hyperlinks to connect their social groups might be considered as a very specific form of bridging which exists only in order to mobilize a social group to act as back-up. It is something more like 'rallying': creating a temporary bridge intended to get people across to other social worlds only for the duration of the argument – a few hours, perhaps a couple of days at most. These bridges between social groups are, in a temporal sense, 'burned' when the rally is over.

It is important to distinguishing this practice of rallying from 'trolling' – it doesn't have the 'disruptive intent' nor the aim to provoke that is characteristic of that activity (Chandler and Munday 2016b). There were certainly examples of nuanced and detailed replies that were, presumably, crafted over hours, and which reflected well-intentioned, critical engagement with the subject matter. But social media activity has very limited capacity to bridge opposing viewpoints when issues are, as in this instance, highly divisive (Hendriks, Duus and Ercan 2016). Even when well-argued, the purpose of much of the argument is not to convince the other side, but to create a space for rehearsing and reiterating beliefs, demonstrating solidarity and doing political performance in a public sphere.

In concluding my interview with the Leeds band, I asked them whether they thought DIY ought to be concerned with bringing about political changes beyond the borders of the scene:

> P4: You can influence people around you to a certain extent but then it's about like, doing it for yourself, and being happy and comfortable [...] I think especially when you realize how terrible the world is, when you come to that age and you're like 'fuck, actually everything's pretty shit', it's nice to have a group around you with the same ideals who all think it's shit.'

P5:   Totally. Or, I feel like a central element of the DIY scene is a community built around safe spaces, so I guess it's just looking out for the people who are there.
P4:   And a lot of the time it's people who don't feel safe in other spaces.

By taking on responsibility for maintaining safer spaces within the limited temporal and geographical confines of the gig, they disavow their responsibility for those outside of it. Quite understandably, their primary concern is for the happiness, safety and comfort of those within the scene, and particularly those who might be at the sharper end of intolerance and prejudice. The work of the DIY bands in denouncing and then debating the pub rock band is best understood as a performance of solidarity, made consciously public through the decision to post updates rather than continue in private conversation with the promoter. The resistant value comes not from engaging with the political other to win hearts and minds, but from reinforcing their own position and displaying their own specific inclusivity and exclusivities – openness as a work of defending borders.

The conflict also allowed for the DIY bands to define themselves as oppositional and other to what they perceived to be more 'mainstream' rock norms. When a member of the pub rock band referred to one of the DIY bands using a derogatory description of their aesthetic, the Leeds band co-opted the phrase, using it as a description under the 'About' section of their band Facebook Page. The band were able to draw on concepts of authenticity from indie-pop and post-punk (Dolan 2010, Kruse 1993) to understand themselves as different to the more masculinist hard rock of their rivals. The value of their conflict therefore ought to be seen as one of restating and reaffirming their principles and, for a young band making some of their first steps in DIY and queer culture, an opportunity to recognize their own identity through recognizing what they were not.

So, while online arguments like this are, in a sense, outward-looking engagements with a wider public, their most significant consequences relate to internal deliberation in the scene. Such arguments are used as a means of affirming identity, and practitioners are pessimistic about the possibility to achieve wider change. Having said that, the DIY musicians' argumentative strategies may have had a more significant impact than they had believed to be possible. Within three months, the pub rock band, who had initially been very willing to defend their decision, were on Facebook advertising their last gig before 'starting up again [...] under a new name'.

6

The popular

Metrics, measurements and the DIY imagination

In the past three chapters we have considered social media's impact on DIY music at increasingly broad social 'levels' – focusing first on the individual practitioner, next on interactions within DIY scenes and then looking horizontally to consider connections and borders between DIY and other local music scenes. This chapter continues that outward expansion: it explores how social media shapes DIY's interactions with the wider world of popular music, and with trans-local audiences. It does so primarily through an examination of the role of social media metrics.

Metrics – quantitative systems of measurement, calculation and quantification – are a key feature of social media. They also loom large in the popular imaginary of what social media *is* and what it *does*. The 'Like' button has been widely (and sometimes glibly) invoked to symbolize the reductive nature of social media, as a tool which purportedly boils down the sweeping gamut of human emotion into a one or a zero: we either Like or we don't. It is through the Like button, argue new media scholars Carolin Gerlitz and Anne Helmond, that Facebook (both within and beyond the borders of its platform) turns 'social interactivity and user affects […] into valuable consumer data' (2013: 1349). Benjamin Grosser links the prevalence of numbers on platforms to a 'business ontology' and 'audit culture' on social media, reflecting on their capacity to provide comparative measurements, and their connection to capitalism's 'growth fetish' (2014).

Setting aside this critical perspective for the moment, it is clear that the growth of social media has brought about the 'datafication' of vast swathes of everyday interaction (Mayer-Schönberger and Cukier 2013: 73–97). Our actions are turned into numbers in three senses: first in their transformation into code, second in their retention as 'data' by platforms and third in their presentation back to us as metrics. This datafication is bringing about new ways of comprehending ourselves and the world around us, such as the rise of the 'quantified self', which makes use of the data points provided by new digital technologies to perform highly metric-based forms of self-evaluation (Ruckenstein and Pantzar 2017). The process of 'making the web social', argues José van Dijck, is also the process of 'making sociality technical' (2013: 12).

Historically, DIY scenes have valued a qualitative approach, as emphasized in the de-massification strategies of post punk and the epistolary intimacy of riot grrrl (see Chapter 2), in contrast to the popular music industries' apparent quantitative focus on sales figures, demographics and profits. Indeed, a focus on charts, sales figures and so on has historically been a feature that has served to define the 'mainstream' in contrast to alternative scenes' pursuit of other, less measurable goals (Anand 2005). The relative insularity of DIY scenes has been associated with the capacity to value quality over quantity, and to find specific meaning in music that is apparently lost when exchangeability is prioritized.

Quantification, in contrast, seems to be entangled in important ways with processes of commodification and reification. While not all quantities are commodities, and countability is not the same as exchangeability on the market, quantitative measurements in general tend to render things comparable at the cost of a loss of detailed understanding of those things. For mid-century critical theorists Horkheimer and Adorno, this propensity of quantification to flatten understanding was fundamental to what they saw as the 'irrationality' of the Enlightenment project's desire for rationality. They saw this epistemology of *reduction* as abetting instrumental reason's complicit relationship to power and domination (and ultimately to fascism), creating an approach in which 'anything which cannot be resolved into numbers, and ultimately one, is illusion; modern positivism consigns it to poetry' (Horkheimer and Adorno 2002). This also highlights the special potential for art to represent an understanding *beyond* numbers (albeit often from a position of irrelevancy) – Adorno recalls that when asked to 'measure culture' he reflected that 'culture might be precisely that condition that excludes a mentality capable of measuring it' (2005: 223).

So the dichotomy, crudely expressed, is that DIY might be notably numbers-averse and that social media is quite famously chock full of numbers. My exploration of this issue is to some extent a development of the scholarship quoted above, wherein metrics are problematically capitalistic. However, rather than seeing 'the numbers' themselves as the problem, I suggest that we ought to focus specifically on what platforms seek to do with them, and how they might attempt to attribute particular ethical and social values to numbers. To that end, I conduct a platform analysis of Facebook Pages which suggests that metrics are indeed in this context deeply imbricated in a harmfully individualistic 'enterprise' discourse.

But to what extent is that discourse then taken on board by DIY musicians? In the rest of the chapter I answer that question, whilst retaining an emphasis on DIY's relationship to 'the popular'. I show that metrics are often valuable to DIY in surprising ways. Firstly, they are useful in constructing an 'imagined audience' online, helping to create meaningful experiences in place of largely absent qualitative feedback. Secondly, DIY musicians are able to connect metrics back to local, material experiences in their scene. This means that the numbers here might carry an important, albeit compromised, ethical meaning – they are certainly not all about gauging popularity.

In other senses the metrics have less positive consequences, notably in the new ways in which they mediate DIY's relationship with mainstream popular music. Again, metrics are nothing without existing discourses within DIY, and I identify a shift towards a politics of recognition as playing an important role here. In this context, metrics allow new kinds of direct comparability, and I argue that the consequence for DIY musicians is often a rationalistic, realist perspective that is unhelpful in maintaining what we might call the 'DIY imaginary'.

I conclude the chapter on a more optimistic note, by considering the limitations on the impact of metrics in DIY. I note that there are still plenty of spaces (both physically and psychologically speaking) where metrics and considerations of popularity hold minimal sway, and that these spaces seem particularly beneficial to practitioners' well-being.

Metrics and platform discourse

♪ 'When numbers get serious' – **Paul Simon**

In order to consider the role of metrics it is necessary to ask: to what ends are they employed by platforms? What kind of social logics seem to be implicated in a specific application of metrics? I apply these questions here to Facebook Pages and demonstrate this platforms' pertinence to studying contemporary DIY music. Findings here are assembled from ethnographic material, as well as an application of the 'app walkthrough' method (Light et al. 2018). I also draw on my own experiences of using the platform as a DIY practitioner.

Facebook Pages is a part of the Facebook platform intended to be used by organizations, businesses, causes and other projects. A Page doesn't necessarily imply a collective responsibility, but does imply that it represents something other than just a person (i.e. it might be for an individual's activity as a comedian, or a tree surgeon or even as a social media 'brand'). Pages offers a public-facing profile which is broadly similar to an individual Facebook user's in appearance and functionality, and similarly also offers a private messaging system, although Facebook Pages offers additional business-oriented features such as auto-reply. Unlike the two-way 'Friends' system of profiles, Pages' connection to users is measured in 'Likes' – the amount of users who have, in effect, subscribed to receive content from this Page on their timeline (i.e. news feed). The 'reach' of a Page is approximately determined by the number of Likes it has, plus any additional 'shares' that content receives, which would bring it to the attention of the network of the sharer. Pages is free to use, but owners can pay to Boost their content, meaning it has a better chance of reaching their existing audience (i.e. those who currently Like the Page). In a more conventional form of advertising, they can also pay to have their content appear in the timelines of users who don't currently Like the Page. Whilst much early Facebook activity around brands and businesses took place in Facebook Groups (see Chapter 5), this has been phased out over the last few years. Pages offer a higher degree of official legitimacy (with a 'blue tick' verification system, similar to Twitter) and primarily support one-to-many broadcasting (and exponential virality), rather than the peer-to-peer (and contained) communication of Groups.

As noted above, all the DIY musicians I interviewed had some administrative control (as either sole or joint owner) over at least one Facebook Page, and virtually all of them were able to tell me the amount of Likes on their Page (and usually their amount of Twitter followers), with a margin of error of one or two Likes. Usually they had seen the numbers within the last couple of days, either because the number was in an attention-grabbing position on the site – 'you see it every time' (P27) – or through receiving Facebook notifications encouraging the setting and hitting of targets: 'I only know 'cos there was something on the page where it's like "invite everyone, get to one hundred people!" a couple of weeks ago, just irking you to pay for an advert or something' (P14).

Facebook has put significant effort into establishing Page Likes (and the accompanying measure of 'reach') as an important measurement of success. Facebook Page admins regularly receive notifications 'from' Facebook, beginning with the notification on a new Page that tells you to Like your own page, in order to make it look better to a visitor. The inference is to imagine one's self as being observed, offering reflexivity in place of the absent audience. Page admins receive notifications instructing them to set an initial 'achievable goal' of 50 Likes. They receive regular reminders about approaching milestones ('You're close to 100 Likes!'), although there seems to be no past-tense equivalent (i.e. 'You've reached 100 Likes!') – the affect inspired by forward-looking aspiration is perhaps more beneficial to the platform than celebratory stock-taking.

Administrators of Facebook Pages have access to a large amount of data on the performance of their Page, contained under the heading of 'Insights'. Here, Page owners can find data on the amount of user engagements (Likes, comments, shares) over specific periods of time (day, week, month, quarter, year), information on the demographics of users who have Liked the Page (gender, location, age), and can also track the total amount of Likes for the Page over time. One feature of Insights offers a table with which to compare one's own Page performance with that of several others. This table comes pre-loaded with suggested comparisons of who to compare with, which generally are Pages within a similar field, with a similar number of Likes. The points of comparison are quantitative – levels of post engagement (i.e. amount of Likes, Comments, Shares and views), recent activity (i.e. posts and content uploaded by the Page owner), total number of Page Likes. On my own band's Page, when I first came across this table, it was pre-loaded with three bands, all of whom were personal friends or acquaintances. Facebook Pages had, I suppose, correctly identified us as operating in the same market – these bands are in some sense my closest competitors, even if I don't see them that way – and therefore it would be hard to fault the algorithm, which is presumably based on our bands' network proximity. Nonetheless to see them presented in this context was rather disarming. The algorithm isn't wrong, and the data it uses isn't incorrect, but the discourse is substantially at odds with DIY's moral emphasis on non-competitive community building.

Features like this comparison table, as well as the aspirational notifications discussed above, show that the discourse of Facebook Pages is normatively oriented towards growth and competition. Even though Pages are not designed exclusively for business use, they encourage an entrepreneurial sensibility, framed so as to generate sustained participation in the platform's internal market for attention.

In this sense tools like Facebook Pages contribute to a growing pressure for DIY practitioners to think in terms of enterprise – even if they aren't actually seeking to make a living from music. In keeping with the spirit of this 'enterprise discourse' (see Banks 2007; 2010; McRobbie 2002),

Facebook Pages shows users the rewards on offer for conformity, but leaves the actual work of self-governance up to individuals.

Into the unknown: Metrics and context collapse

Social media scholars Alice E. Marwick and danah boyd have written on the prevalence of 'context collapse' on social media (Marwick and boyd 2010), drawing on the work of Erving Goffman to consider how our performance of identity varies significantly across our different social worlds in order to remain context-appropriate. They argue that social media 'collapses' these separate contexts (i.e. our family, friends, colleagues, associates and potentially strangers), creating one space in which we have to communicate a single message to multiple audiences. The result is an increased emphasis on the 'imagined audience', a means by which social media users attempt to avoid communication failure by imagining who *might* be on the receiving end (Litt and Hargittai 2016).

For DIY music practitioners, the work of 'imagining' an audience online is often difficult. At times, this means acknowledging that the digital traces left by audiences are open to multiple, contradictory readings, as in this consideration of Bandcamp download stats: 'Our demo has had 21 downloads, 'cos I looked a couple of days ago. And it's like, who are those 21 people? I'd like to find out who has downloaded it, because I guess it's just 21 people I know. It might just have been [our drummer] 21 times' (P14). This is in stark contrast with experiences of playing shows at local DIY venues which, as I showed in the previous chapter, are highly predictable in terms of audience, and where the band might feasibly know every person in the room. This reflects the specific ability of the internet to promise the improbable through its networked capacity ('anyone could see it!'), whilst denying that possibility as a result of the very same capacity ('there's too many people – what chance do I have?'). Laura Gurak identifies 'speed' and 'reach' as key features of internet communication that act as complementary 'partners' (2001: 30); Nancy Baym uses Gurak's terms to distinguish that a key feature of social media is its ability to make digital texts travel far and wide and fast (2015: 12). This is true insofar as they are *capable* of reaching a global audience within seconds, but in reality, particularly at this level of obscurity, in a type of scene characterized by close-knit networks, this kind of unpredicted travel seldom occurs. Belief in this kind of mythic internet, which is supported by the exceptions (i.e. the viral sensations) rather than the majority of unsensational content, only obscures our understanding of how information flows through networks. The unknowable audience of the internet offers an updated version of the 'talent scout in the crowd', presenting the possibility of mysterious strangers in the dimly lit audience; the likelihood is that, should the lights go up, they will reveal the expected familiar faces.

In the context of this uncertainty, and the difficulty of accessing reliable qualitative information, numbers such as Likes on Facebook Pages become central to imagining an audience. Likes carry validity because they are longstanding, widely used and publicly comparable (unlike, for example, Bandcamp download stats, bands can see *each other's* number of Page Likes). Contemporary DIY musicians increasingly rely on quantitative metrics in order to imagine their audience largely because qualitative understandings of online audiences are difficult to reach. The 'imagined audience' – this device through which we reconcile reality and our expectations – is most easily imagined through the conduit of metrics.

Making sense of metrics

Nancy Baym argues that quantitative metrics fail to 'see' or 'capture' data in two significant ways (2013). The first problem is that metrics are inaccurate, sometimes deceptively so. This is largely because they fail to filter out interactions from bots and other non-relevant actors (a failing reported widely in mainstream media, for example, Cellan-Jones 2012). The second is that they fail to capture depth of feeling or the affective dimension of the interaction that they record. The question of whether metrics fail is critical here insofar as it might highlight their limited capacity to support the 'enterprise discourse' outlined above, and thereby allow for other meanings to be made and found through metrics.

In terms of accuracy, DIY musicians held some reservations, but none strong enough to consider the measurement of Facebook Page Likes as fundamentally 'broken'. In this example, 'inaccuracy' doesn't mean recording the activity of bots, but rather the activity of close friends:

> Half of it is probably just my friends I guess. That's not that many people. And sometimes I'm like 'I wish I could just start again', and start with zero. And anyone who actually likes the music just like it, rather than having loads of friends who like everything all the time. Cos I'm just like, 'your like isn't really worth anything', that's not an accurate gauge of if people like my music or not. And I find that pretty irritating sometimes. (P15)

In this description, the (impossible) aim of the metric is to record the music's *objective value*, but its application is skewed by social ties that generate good-will towards friends' creative projects. In general, though, the metric was considered to be broadly accurate. One practitioner was similarly emphasizing its shortcomings as a measurement tool, before concluding: 'But I think yes, if it went up suddenly by a thousand, we'd all be like "oh my god, likes on Facebook, this equates to people liking us"' (P14). The numbers are understood to *broadly* correlate to audience size. The metric, as a tool for imagining one's audience, is satisfactory. (In any case, the accuracy and applicability of the metric are to some extent socially determined by others – it matters because promoters say it matters.)

In terms of the realm of affect, it is true that this metric is not adequate for showing, as one practitioner put it, 'the quality of Like' (P15) – that is, the depth of feeling behind the decision to click that Like button, which might be half-hearted or impassioned. But quantitative metrics might still have an affective dimension, particularly for the 'owner' of the metric. One practitioner spoke on a theme that was common in interviews, of Page Likes as symbolizing a general show of support: 'I know that we've got about 200 likes on Facebook page for [our project] and I'm like, oh that's quite cool, 200 people like what we're doing, that's really nice. So I'm not embarrassed about it at all, I see it as a cool sign that people are interested in the stuff that I'm creating' (P9).

So, the metric of Facebook Pages Likes does not fail on Baym's key areas of sufficient accuracy and recording affect: DIY practitioners in Leeds tended to see the metric as valid and found meaning in it. This meaning is often specifically linked back to the local DIY scene and to live performance:

> It's nice to see it [i.e. the number of Likes] go up, cos it's usually when we've played a gig, so that means that we did a good job of playing a live gig, that people have gone on afterwards and want to follow us. (P16)

> That's quite nice to be like, hey, I remember I saw that person last night, they've now come and found us and Liked us on Facebook 'cos they obviously enjoyed the set. I like getting new Likes because you can relate it to those moments. (P27)

A fundamental aspect of Likes is that they are the outcome of a process of reification; they are social relations made into a thing that seems to operate with its own force (and, slightly confusingly, that 'thing' is also immaterial). This reification is compounded by the tendency for social media 'texts' to hang around – what Baym calls storage (2015: 7) and boyd calls persistence (2013: 11) – and to be looked at whenever the user desires. In being reified, they take on a value separate to the social relations that they are intended to reflect. Grosser considers this in terms of Baudrillard's concept of the 'simulacrum', as a sign with no referent (2014). This is even implicitly acknowledged by Facebook, in their invocation to like your own Page because it looks good to others, thus demonstrating that Likes might signify something other than an individual's 'liking' for the thing in question.

Bolin and Schwarz argue that social media data often presents information at a level that is too abstract to be meaningful, and which therefore generally needs to be 'translated back' into more traditional categories of understanding. They find evidence of this happening in news media at both the organizational and individual level (2015: 8). The two interview excerpts above suggest that something similar happens when DIY practitioners find meaning in reified metrics – they translate the quantitative back into something qualitative. The meaning of the numbers here comes from practitioners' capacity to unpick the reified numerical representation offered to them, in order to review and retain the social relations that went into it.

The notion of 'mining' social media data is increasingly commonplace, and Helen Kennedy identifies this activity taking place in 'ordinary' small organizations as well as global corporations (2016). 'Mining' nonetheless feels like an excessively industrial descriptor in this case; I think the activity of unpicking metrics in DIY is more appropriately characterized through the related subterranean metaphor of 'digging'. Digging, like mining, suggests an unearthing of value through processes of excavation, filtration and assessment. But where mining takes place on an industrial scale, with a methodical approach and specialized tools, digging is undertaken by individuals and small groups, scrabbling inefficiently for value amongst the detritus, using simple tools – out of Kennedy's four categories of social media data mining tools, Leeds DIY practitioners used only the simplest, 'in-platform' resources (2016: 20). We frequently 'use new media for interpersonal purposes', says Baym, but since social media offers fewer 'social cues', we 'come up with creative ways to work around barriers, rather than submitting ourselves to a context- and emotion-free communication experience' (2015: 64). Digging for meaning within metrics operates in this way. It is an affordance made possible by social media's tendency to offer persistent and explorable metrics, and utilized by practitioners who have access to the data, but who have limited conceptual and technological means of interpreting it, and who seek to apply their own moral frameworks to their findings, rather than to submit entirely to an entrepreneurial logic.

♪ 'Digging for something' – **Superchunk**

I also want to consider this social media 'digging' as a kind of playful 'pottering around' with data and metrics, which often operates in a way rather different to that intended by the platform. For example, one practitioner reflected on the usefulness of having demographic information on their Page that informs them where in the world their Likes are coming from:

> It's vaguely interesting, I guess? Like, that they break down every city and country, the people that like you ... that's interesting. But, again, it's sort of indifferent because it doesn't make any difference to ... I don't mind, or, I don't care who likes it, it's just nice that they like it. I really don't mind who it is, their age or their gender or where they live. It's completely irrelevant, basically. (P19)

With no marketing plan or global expansion strategy (due to scale, perhaps, but also an aversion to such things), this data can't be instrumentalized or acted upon, rendering it interesting only in an abstract sense. Finding this data 'vaguely interesting' constitutes a sort of non-participatory resistance; even if metrics audit for growth, they might be counteracted without much effort by lack of interest in competition, or an unwillingness to distinguish people based on national borders.

So, practitioners demonstrate a fruitful ability to look beyond metrics and back into the social relations that form them, and to ignore and misuse data. Arguably this affective dimension only comes about because Facebook specifically allows Page owners to delve into that information, and see the names and profiles of individual audience members who constitute the whole. But from here it is also possible to see how the number as a whole might continue to serve an affective purpose, as a representation of accumulated moments of memorable sociality. This is beneficial, inasmuch as it suggests that an engagement with metrics needn't constitute the loss of the specifically emotional connections associated with music (see Jakobsson 2010).

The ability of the number to reflect values other than growth also carries a moral dimension. One of Facebook's key incentives to encouraging an aspirational approach amongst Page owners is to encourage the usage of Sponsored Posts – advertisements that cost little (starting at £3) and can 'boost' the reach of a post. There is a significant incentive to do this, as Facebook tightly restricts Pages' capacity to reach audiences without paying (and seems to be increasingly tightening the squeeze on 'organic', i.e. unpaid, reach (Loten, Janofsky and Albergotti 2014)). However, several Leeds DIY practitioners suggested that to pay Facebook for this service would be unethical: 'It doesn't seem organic to me. That doesn't seem grassroots or DIY to me, to pay Mark Zuckerberg some money to post our advert. I dunno, maybe I'd rather do a whole fucking other tour to get the exposure that that Facebook post would get' (P16).

Another practitioner similarly advised: 'You've gotta earn your Facebook Likes. You've gotta gig and earn them' (P4). This was delivered slightly tongue-in-cheek, perhaps aware of the extent to which this rhetoric echoed more conservative notions of hard work bringing about reward. But nonetheless, part of the reasoning behind refusing Sponsored Posts is about keeping that number of Likes *honest*. Marx drew upon the Roman Emperor Vespasian's declaration, *pecunia non olet* ('money has no smell'), to expose money as the only commodity which, having only exchange-value and no specific use-value, was able to fully hide its origins (Marx 1976). But some Facebook Likes, apparently, do 'smell' a little off. This is to say, in Marx's terms, that something of the use-value lingers and affects the extent to which Likes can offer an untainted exchangeability or act as a transparent means of comparison between bands.

The beneficial consequence for DIY ethics here is that Facebook Page Likes might carry a metric measuring something other than growth. A quantitative metric, designed only to measure and compare size in positive terms (i.e. the bigger, the better), might be subverted in order to represent a degree of ethical purity or distance from commercialism. This might provide a means of resisting the 'self-blaming' that is central to enterprise discourse, seeing smallness not as a lack of fortune or talent, but as a manifest consequence of a specific ethical approach.

But this quality is undermined in a couple of ways. Firstly, it's not always clear what's going on in the murky world(s) of music promotion and online marketing (e.g. the ambiguous rewards offered to social media 'influencers'), and therefore not at all obvious which bands have earned their Likes by fair means or foul. This means that *it isn't a very good measure of ethical value*, even if some practitioners use it to this end. Secondly, many DIY practitioners have a stubbornly persistent belief in a positive correlation between talent and reward, best summarized by the maxim that 'cream eventually rises to the top'. This perception that the smoke-and-mirrors of marketing 'can only do so much' means that whilst a band's large audience might be to some extent 'inauthentic', it is also to some extent deserved. Fame and success are self-legitimating, and by implication, so are obscurity and failure. This leads us on to another dimension of metrics in DIY – their role as mediator for the relationship between DIY and popular music.

Popular music and the DIY imagination

Metrics are, of course, not the only variable in DIY's changing relationship to the popular. I wish to suggest that they play a quite specific mediating role, which I outline below, but that this role cannot be understood minus the relevant socio-historical context.

There has been a major shift within 'alternative' music circles in the twenty-first century, towards a populist discourse sometimes summarized as 'poptimism'. This term was initially used to describe a school of music criticism (on sites like Pitchfork and the now-defunct Stylus Magazine) which took the work of chart music producers seriously and treated the multi-faceted contributions of pop music artists as meaningful beyond a focus on whether they did or did not involve themselves in the work of musical composition (Rosen 2006). Whilst the term itself isn't in everyday use, the approach is unquestionably dominant now, and has reach beyond the realm of music criticism. DIY has always had an ambivalent relationship to popular music (as I outline in Chapter 2) and, more than that, alternative and indie scenes have always had their favourite pop stars – inspirations who have acted as both precursors and parallels to their own music-making. But today this interest in mainstream popular music is more fully-fledged; it constitutes a political faith in the capacity of some mainstream pop artists to advance goals of social justice – and to have fun doing it. The sense of pop as something enjoyed ironically, or as a 'guilty pleasure', is certainly entirely absent. Poptimism was not the first attempt to problematize the exclusions and omissions evident in constructions of rock authenticity (which, as explored in Chapter 2, overlaps substantially with DIY authenticity) – it strongly resembles a similar concern over 'rockism' in the 1980s UK music press – but in the present day guitar-based music holds an even weaker claim to give voice to social discontentment. Pet Shop Boys' vocalist Neil Tennant, a veteran of this earlier ideological skirmish, recalls that the 'whole thing [i.e. debate] went away … because rock lost' (Needham 2018).[1]

This connects to changes touched upon in previous chapters. It certainly connects to the fact that, as explored in Chapter 3, the 'intimate' communicative style that was previously a hallmark of DIY is now a requirement for all kinds of musicians on social media – celebrities appear shorn of whatever machinations might be working to sustain their renown. The shift towards a politics of visibility (further discussed in Chapters 3 and 7) is also critical. If the aim is to represent particular social groups through being visible then, generally speaking, the more famous the artist, the better.

In this chapter there is limited scope for exploring the full extent of these discursive shifts, but this brief overview has I hope made it clear that they are significant. What is of interest here is how metrics relate to or reflect these populist discourses in particular ways. Metrics make it increasingly difficult to locate and elucidate *qualitative* differences between DIY and popular music. They make it very easy to locate *quantitative* differences, and they assign greater value to larger numbers.

When the primary means of 'measuring' one's own impact are quantitative, and therefore all too easily comparable to the user-engagement levels enjoyed by mainstream music stars, a qualitatively informed conception

---

[1] This raises the fascinating question of why so many DIY practitioners play rock music when they have very little faith in the genre's political potential.

of DIY's value (whether aesthetic, ethical or political) becomes harder to sustain: 'I think it's a bit weird to be like, oh, my music, that no-one's ever gonna hear, is really, really good, and is way better than Drake, even though twenty-million people watched his video. It's just like, it doesn't make any sense, that doesn't make any sense whatsoever' (P12).

The word 'sense' in the above quote is revealing – the ideological dominance of exchangeability is understood here as undeniable common knowledge. Metrics are powerful sense-making tools which, for all their malleability, prioritize the rational. They show a particular reality which can be hard to disavow. That particular reality is not helpful for DIY's sense of itself as politically meaningful. In the previous chapter, playing at Leeds' DIY venue Wharf Chambers made at least one practitioner feel 'famous'. Temporarily setting aside the problematic, potentially narcissistic dimensions of fame here, we can say that metric understandings of audiences online ensure that DIY practitioners feel resolutely *un-famous*. A significant downside of this quantitative understanding of online audiences, then, is that an 'imaginative' dimension of DIY is negated.

♪ 'Nobody' – Mitski

Imagination is important to the political value of DIY music. It is notable that DIY music, unlike many other forms of participatory amateur music-making, is premised on the re-creating of the music industries in microcosm: centring around DIY record labels, DIY bands, DIY tours with DIY merchandize and so on. As I outlined in Chapter 2, the communicative model proposed in DIY is essentially that of the pop or rock star, and as such involves a collective suspension of disbelief, in order to posit this small-scale activity as culturally significant mass communication. Implicit in the 'anyone can do it' mantra of DIY is a speculative assumption that we might all be able to play at being a pop star. Imagining, in DIY – or, more prosaically, 'pretending' – is an important part of its capacity to offer self-realization and a means by which it offers a 'voice' to its often-marginalized practitioners. It is part of the process that might allow a DIY musician to believe, in whatever ethical or musical sense, that they are the equal of a best-selling artist like Drake.

Social media platforms aim, in general, to disavow the imaginary, since their business models increasingly rely on real world 'embeddedness'; 'Real Name' policies on platforms, for example, restrict the use of non-birth names in order to demonstrate to advertisers that user profiles are indeed 'real' consumers-in-waiting (Gerrard 2017). And whilst social media metrics might be playful insofar as they allow a certain 'gamification' of social and cultural practices (Whitson 2013), they are not helpful in sustaining the sense of playfulness that allows us to imagine the world as other than it is.² Metrics on the whole assume, and perhaps demand, a serious intent – particularly when attached to the entrepreneurial, individualizing discourses of platforms such as Facebook Pages. The 'imaginary' dimension of social media is countered not only by measurability, but also by the temporal framing of what is measured. Metrics are focused on the capturing, solidifying and recirculating of social interactions that have *already happened*, to be used as the basis for predicting and limiting future interactions. Their status as quantified measurements gives them a particular power of reification.

As I have demonstrated in this chapter and elsewhere, DIY practitioners have a notably self-depreciating perspective on the value their music might have out in the wider world beyond their scene. There are other factors at work here, beyond (or alongside) the impact of social media. In genre terms, there is a steadily declining belief that guitar-based music might act as a meaningful communicator of social change, replaced by a faith that popular music might carry the hopes of an increasingly recognition-based political project. And, as noted in Chapters 4 and 5, the offline intentions of DIY practitioners in manifesting an insular scene reflect an understandable and often valuable desire for safer spaces. But I've suggested here that this desire for insularity interacts with particular forms of data presentation on platforms and has consequences that are perhaps less valuable. Platform metrics show that DIY audiences are undeniably small and as such invite comparison with more clearly 'impactful' pop musicians. Metrics do not fully determine the nature of this comparison, but they combine with prominent discourses to make particular kinds of measurements. Important political dimensions of DIY might be invalidated by these measurements and the modes of interpretation that they propagate.

Forgetting the numbers

♪ 'Carry the zero' – **Built To Spill**

So, metrics play a large role in the interactions between DIY and popular music by reifying musical and communicative activity in a specific way. They

---

²Julie E. Cohen has written on the ways that the 'networked self' might relate to themes of creativity and identity, suggesting that the ways in which culture and subjectivity emerge are often related to playfulness in everyday practice (2012). Cohen is particularly concerned with copyright regulation online, but also points more generally towards the potential for the internet to, through both its architecture and its discourse, constrain or enable imaginative dimensions of our lives. She argues powerfully that 'the design of information architectures should be guided by the need to preserve room for play in the use of cultural resources, in the performance of identity, and in the ongoing adaptation of places and artefacts to everyday needs' (Cohen 2012: 224).

help to put popularity front and centre as the means of determining value and political impact. But an important question concerns the extent of their powers: how far do metrics reach into the offline manifestations of DIY activity? How prominent a role do they play in DIY musicians' psychic lives? Here, a practitioner reflecting on the (in their eyes) excessive focus on metrics maintained by a friend then turned towards their own practice and their changing approach to the numbers:

> I found out the other day that [my fellow band member's] old band would delete a [Facebook] status if it got less than something like 12 Likes. They'd say 'it wasn't popular, delete it', cos how does that look to a label, sort of thing. And that's just such a conceited, contrived way of living your online life. I'm not saying that's [not] something that I maybe have thought in the past – that didn't go down well, so how does that come across to people, or do we look lame that we said something kind of stupid, but I just don't give a shit anymore, I don't. That's such an empty way of living, it's not colourful … you're equating things all the time, you're considering things so much, when you could be chilling out and having a nice time. Rather than worrying who did or didn't like a status. But the internet can make people neurotic like that cos it's a competition. (P16)

This practitioner reiterated a few times during our interview that they were consciously trying not to think about the numbers, in order to enjoy the experiences of writing and playing music more fully. As described above, practitioners do frequently find positive values in metrics shown to them. But, given their inescapability, and the extent to which they drive feelings of insecurity, there was a general sense that overall it would be better not to know.

When DIY scenes assign meanings to metrics, then, there is a duality of perspective at work. On the one hand, quantitative measures reinforce feelings that the fame and fortunes of popular musicians are both deserved and valuable, and that metric growth constitutes success. On the other hand, practitioners know that there is some meaningful dimension of their activity that remains uncaptured and that Facebook Pages measurements are 'shallow'. One practitioner directly equated this ambiguity towards their Likes with a feeling of switching between two distinct mental states:

> I go in and out about caring about Facebook Likes, and when I'm more insecure about music, that's when I know all my numbers. And when I'm more secure about my music that's when I don't care, about anything. […] I think it's just recently with the release of that album I was just very emotional and temperamental about everything, so I know all the stats for everything that ever happened. Whereas also like, the numbers really don't mean anything. They really, really don't. […] That's something where there's loads of cognitive dissonance in my brain about it. (P15)

Reification here has the beneficial consequence of the numbers sticking around, echoing the 'persistence' that danah boyd has identified as one of the four key affordances of social media (2013: 11). This means that when this practitioner is feeling insecure, the numbers carry the highly valuable sense of actually being there. They reflect the undeniability of the fact that *something happened*, even as they alienate the practitioner from their feeling of creating or controlling that activity. However, it is important to understand the deeper consequences of seeing reification as security. Finding one's own worth in the exchangeable (i.e. I *am* 500 Likes ... and that's *something*) is an act of self-commodification, a championing of exchange value over subjectivity. Through this frame it is difficult to value one's self other than through one's mediated representation – identity here is, as Jodi Dean puts it, 'reflexivity that goes all the way down' (2010).

On the other hand, as the interview quote above demonstrates, there are times when practitioners feel 'close' to the music and when the numbers seem not to interfere. This might be considered in terms of 'flow' (Csikszentmihalyi 1990), where a high degree of immersion in musical practice results in a strong sense of the self as unified and embodied. There are specific times when this flow is in operation – during writing, rehearsing, playing and performing. Flow is often afforded during musical activities that are 'clearly bounded by time and space' (Turino 2008: 5), and in this context the un-ending, infinitely scrollable distraction of social media is a significant barrier to immersion. Flow is also linked with a loss of self-consciousness (Hesmondhalgh 2013: 33), where one is not reflecting on how one appears to others. The construction of the self is far more secure and seems to be enacted with no barrier between thought and action, or between meaning and intent. The nature of these moments is such that it is difficult for metrics to get 'in-between' practitioners and their musical engagement. These are areas that are currently beyond the 'capture' of communicative capitalism. When the 'flow' is good, it is not that the numbers take on a different, *ethical* meaning, but rather that they cease to mean anything at all.

Katharine Silbaugh, writing on the rather different subject of whether academic exams constitute commodification, argues that what is so destructive about quantification is that it 'places pressure to re-design the world so that we place our energies behind only what is measured' (2011: 325). This highlights the way that power is exercised not through measuring the success or failure of individuals, but through the putting in place of specific measurements that then work to shape individuals' and groups' sense of what is valuable. Those in control of measurements are in control of what is audited, and we work to meet their criteria. In this case platforms have a high degree of power which is exerted not only through the metrics themselves, but also through the enterprise discourse presented in notifications and other features, which aims to accelerate content circulation. However, as Jurgensen's 'omniopticon' model of social media suggests, the *social* is still fundamental in how meaning is constructed through metrics (even though, for Jurgensen, this is sociality put to work as surveillance) (2010). Dawn Nafus argues that as data becomes 'domesticated', people develop their own 'sense-making methods' in a variety of ways (2016: 384). Even in an 'audit culture', we have some say over the values that we hold dear, and they may not be the values that are measured.

This chapter has shown that metrics and datafication do turn DIY practitioners' attention towards concerns over popularity in pernicious ways but, in two significant ways, this influence is limited. Firstly, practitioners are capable of 'digging' to find different meanings in quantitative metrics: they emphasize qualitative dimensions of metrics and are sometimes able to apply their own moral frameworks to the numbers provided. Secondly, there are still plenty of times and places where people are able to forget about the metrics and focus on rewarding and satisfying activities (such as music-making), even as other dimensions of these activities (i.e. the reification of music activity into social media metrics) exacerbate competition and self-commodification.

7

The platform

Alterity and the political economy of social media

In 1980, the London-based DIY band Scritti Politti were given a five-minute segment on BBC Two's 'community action' magazine show *Grapevine*, in order to outline the processes involved in do-it-yourself record-making. In it, they cover the basics of recording a song, mixing it, pressing it to vinyl, creating artwork, and dealing with distribution and promotion. The segment is introduced by the show's presenter Ann Barker, to camera, asking her audience: 'Have you ever thought of bringing out your own record? It sounds impossible: surely it's too sophisticated a process, too commercialized, and too sewn up by the big companies? Well, that may have been the case a few years ago' (Meads 2016).

However, even whilst attempting to demonstrate its relative simplicity, Scritti Politti's record-making guide makes the process look, at least to modern eyes, prohibitively difficult and expensive. Making one-thousand 7" singles takes them around two months, involves 'a lot of time on the telephone', utilizes multiple light-industrial processes and costs around £500 (equivalent to approximately £2,000 today). DIY circa 1980 required a great deal of resourcefulness, not to mention financial wherewithal, in order to surmount still-significant barriers to entry.

In contrast, the equivalent DIY processes today involve significantly less difficulty, and far less financial outlay. This change was in part a gradual process: the development and increasing popularity of the photocopier, tape cassette and home studio technology (Théberge 1997) made record- and zine-making incrementally easier and cheaper in the intervening forty-odd years. But mainly this process has happened rapidly and recently, as a consequence of the personal computer becoming a ubiquitous feature of everyday life for many people, and the internet becoming the primary conduit for finding and listening to musical recordings. Digital production has substantially lowered the cost of recording music, and online distribution platforms mean that sharing recordings with an audience is often completely free. (So, whilst physical releases are still prized, they are no longer necessary in order to get music heard.)

DIY emerged in an age in which doing-it-yourself 'sound[ed] impossible'; it operates now in a world where DIY culture has, according to the rhetoric at least, become commonplace. This book has outlined the benefits that this has brought, but it has also attempted to engage seriously with the ways in which that ubiquity might constitute a 'loss' – and not merely a loss of subcultural cachet. This final analysis chapter considers this question in the context of academic literature on 'platformization'. This literature places online platforms at the centre of ongoing large-scale social, cultural and economic reconfigurations, and views platforms as possibly underpinning a new form of capitalism.

The first section focuses on how platformization might re-configure norms of cultural production and the consequences for DIY music. Building on the historical example of Scritti Politti outlined above, it draws on Harry Braverman's work on 'labour process' in order to consider how and why changes in systems of cultural production (such as platformization) might also change DIY's capacity to offer a meaningful critique of the cultural industries. DIY can be understood as, effectively, a critique of Fordist and Taylorist labour processes in cultural work, and this raises substantial questions regarding its capacity to offer critique of platforms as a post-Fordist 'workplace'. DIY's historical rejection of managerialism is shown to be deeply, unfortunately, compatible with the political-economic logic of platforms.

Literature on platformization has also focused on identifying the 'operational logics' of platforms – that is, how platform discourses create new routines and behaviours within the spheres of culture and society that they touch upon. In the second section of this chapter, I argue that 'optimization' is one such key logic, and one which poses a particular challenge for DIY music. It is a logic by which branding and marketing norms are made part and parcel of social media usage. Optimization argues that since the online consumer has substantial agency to make their own free choice, any work to help them find content does not bear the deleterious social imprint of marketing. It therefore has the potential to make individualistic branding norms commonplace within DIY music (and beyond).

Social media and DIY 'deskilling'

DIY music today is easier than ever. This is not to say DIY music and other kinds of amateur or non-professional music practice have merged entirely. As outlined in Chapter 1, there is a distinction between DIY and what I will call 'platformized' DIY – that is, those aspects of practice which use online platforms and other related digital tools, and which are increasingly undertaken by both DIY and non-DIY practitioners alike. But there are new ambiguities within DIY about what doing it 'yourself' might mean, and whether it carries any meaningful sense of alterity. As one practitioner outlined in our interview:

> A whole load of people are talking about DIY [as] just about literally doing it yourself. Like, what does 'doing it yourself' actually mean? Writing your own letter to the record label? Is it that? Like, driving to meet the record execs yourself, not getting a taxi? The very minimal amount of what you're doing, how independent you are ... yeah, it gets used as the word independent. This is the problem, cos now DIY means something different to what it used to, it used to be a lot more about politics. And what I mean by independence is like, well everyone's independent now cos there's no real major labels so ... you've got to be independent, but that doesn't mean you're DIY in my definition of it. (P12)

This quote, which echoes feelings particularly common amongst older practitioners, suggests that DIY is going through something of a crisis of meaning, in part caused by the 'opening up' of DIY as a consequence of platformization. This connected to a widespread perception that the platformization of DIY was bringing about the loss of some real, practical skills possessed by DIY practitioners (although these specific skills were often hard to define precisely).

The comparative ease of contemporary DIY was acknowledged by Leeds practitioners who remembered pre-internet practice or who had been informed about it by others. One recalled 'the tougher side' of DIY as it was in the 1990s and suggested that younger practitioners might '[not] be willing to go through that now'. Practitioners emphasized a lost materiality that 'was like the most "DIY DIY DIY" you could get, basically'; in contrast, 'just using Facebook' was seen as signifying a lack of effort and commitment. In these ways, a dichotomy of ease and difficulty was used as a shorthand for what activities met a threshold for 'real' DIY.[1]

---

[1] This emphasis on grimy physicality as constituting authenticity can contribute to situations which exclude or marginalize some people with physical disabilities – for example, the punk tradition of the 'basement show'. Advice on improving accessibility for DIY shows can be found in the excellent *DIY Access Guide* (2017), compiled by the disability-led charity Attitude Is Everything.

But such valorization of hardship and difficulty begs the question: isn't DIY *supposed* to be easy? Or, to phrase it in terms more closely aligned with democratic aims, shouldn't we *want* everyone to be able to 'do it themselves'? Isn't there a way that we could extol and defend the virtue of DIY ethics, without recourse to what does and doesn't count as 'real' work?

To begin to answer these questions, we can return to the example of Scritti Politti. The relatively difficult DIY process they outlined in 1980 carried an implicit critique of the cultural industries, which can be placed in historical context with reference to the work of labour process theorist Harry Braverman. Braverman's book *Labour and Monopoly Capital* (1974) is a Marxian analysis of the consequence of 'scientific management' in the workplace (sometimes called Taylorism after its pioneer Frederick W. Taylor). Braverman argues that before Taylorism developed in the twentieth century, most manual labour involved a combination of 'planning' and 'doing', whereby workers thought about the best way to achieve their variety of tasks and then physically carried out that plan. But Taylorism, in a quest for maximal efficiency, encouraged the division of such work into small, repetitive tasks – as in Fordist 'assembly line' production. Workers' 'doing' activity was increasingly divorced from the 'planning', which was performed by a handful of specialist managers. While this served the capitalist imperative to keep production costs low, for Braverman the really significant outcome was 'deskilling': the transferral of 'planning' knowledge from workers to management (1974: 41).

Braverman's key argument was not that workers had lost *specific* craft skills (a popular nostalgic lament), but that capitalists' control of production processes was used to appropriate workers' knowledge and hoard it within a management class. In this context, Scritti Politti's resourcefulness served as a critique of Fordist capitalism and Taylorist management strategies, in which taking oversight over the whole production and distribution process constituted a radical reconnecting of planning and doing.[2] Their pessimistic view of the cultural industries led Scritti Politti to see musicians on major labels as but one cog in the machine; their solution was to posit a system in which artisans retained managerial oversight, rather than allowing their 'doing' skills (in this case, music-making) to be subject to the 'planning' of others. The extent to which music production has ever been Fordist is questionable, since the cultural industries have long been required to accommodate artisanal modes of production (Toynbee 2000, Banks 2007), but nonetheless it was owners and managers who found systematic ways of harnessing this activity, rather than musicians themselves.

---

[2] From the musicians' perspective, that is. The processes documented in the *Grapevine* video clearly still involve skilled manual labour performed by, for example, pressing plant employees – there is a limit to what the band are physically 'doing' themselves.

The platforms that predominate today are capitalist entities, but many of them – particularly those that engage with creative and cultural work – do not rely on Fordist and Taylorist modes of production for the majority of their profit. They seek to capture activity, rather than to produce goods. For Nick Srnicek, platform capitalism is 'centred upon extracting and using a particular kind of raw material: data' (2017b: 39), meaning that platforms do not necessarily own the cultural product itself, but rather they 'provide the basic infrastructure to mediate between different groups' (2017a: 44). In this way they present themselves (somewhat misleadingly) as 'empty spaces for others to interact on' (2017a: 48), whilst gathering data on transactions that can be sold to advertisers.

This in turn brings about an important change relating to efficiency. Bernard Miège's seminal analysis of the economy of cultural production (1989) identified the importance of non-professional practitioners in sustaining the cultural industries. In fields such as popular music, un(der)paid amateur producers functioned similarly to how Marx posited the function of the 'reserve army' of the unemployed (1976: 781–93): as a social force that compels workers to produce efficiently and effectively, else be replaced. But platform capitalism encloses this 'reservoir'. 'Amateur' practitioners of various kinds – DIY musicians, but also YouTubers, photographers, novelists and so on – undertake much of their cultural activity within the confines of the platform (or across a series of platforms). Unlike record labels who, due to limited time and resources, have to choose their roster of artists carefully, platforms need not invest in any specific users. There is, materially, no obvious upper limit on how many practitioners they can offer 'space' to. Consequently, there is no great onus on platforms to push their users to be more efficient, that is, to *manage* them into producing a higher quality or quantity of content, so long as they are getting the data they need – platforms' efficiency concern lies in being able to get more data than their rivals can and to make more of it. (This need for data efficiency does lead to particular organizational logics, explored in the next section.)

What platforms offer us, in return for our circulatory zeal, are tools. Platforms offer a wealth of accessible, automated solutions in order to make it easy for one person (or one group) to take control of processes from start to finish. Bandcamp and SoundCloud take care of global distribution within minutes, and can be reinforced by services like DistroKid and Tunecore that register music to the major streaming and digital sales platforms; Dropbox is used to send files privately to allow near-instant collaboration; WhatsApp and Facebook Messenger provide the communication infrastructure to organize gigs; Facebook Pages and Events are the means of publicizing and marketing. There are countless other, more 'subterranean' platforms that support and augment this digital infrastructure. The holistic approach valued by DIY is now easily enabled by using each platform's automated solutions in order to retain oversight of an entire project.[3]

In Scritti Politti's DIY heyday,[4] it was 'planning' – that is, maintaining a managerial oversight over musical activity – that generally constituted the highly valued side of cultural labour. For musicians today, broadly, the 'planning' is easy. Maintaining oversight, or enacting 'self-management', is the default. It is the 'doing' – the enacting of our plans – that is handed off to automated systems, and which has consequently become unfathomably complicated.

Of course, this planning and doing dichotomy simplifies a system that remains immensely complex.[5] But the shift is significant. DIY practitioners today – and music practitioners of all kinds – have remarkable access to the kind of oversight and autonomy that Scritti Politti and their early DIY cohort were so enthused by. And, despite concerns over loss of material competence, there are new skills being learned by contemporary DIY practitioners too: not only in terms of project management, but also in gaining technological know-how and the ability to traverse new immaterial economies.

But Braverman's work on scientific management comes to the fore here, shedding light on this new economic sector that seems to bypass managerial strategies entirely. Crucially, his understanding of 'deskilling' centres not on attempting to qualitatively compare skills lost and gained, but on the *relative* values of these skills in relation to the wider economy and the impact of deskilling (and new technology) on the division of labour and wages under capitalism. This is distinct from some readings of Marx that would find certain aspects of labour as 'essentially' human, and thus our distance from them as dehumanizing.[6] Deskilling is not about lamenting a loss of specific skills: the important thing is whether the distribution of skills in labour processes tends towards *averaging* value (i.e. creating greater equality in the work force) or whether it *polarizes* those whose time is 'infinitely valuable'

---

[3]There is also some evidence that platforms are moving towards simplifying and automating increasingly more 'creative' territory. For example, LANDR provides automated mastering services, something generally considered to be a highly specialized and *embodied* craft (prized mastering engineers are said to have 'golden ears').

[4]Scritti Politti's musical 'heyday' arguably came later when, after losing faith in the political project of DIY, they released the pioneering synth-pop record *Cupid and Psyche 85* on Virgin Records (1985).

[5]For example, there is a tendency in the cultural industries – even amongst platforms – to seek vertical integration (Hesmondhalgh 2019: 226–9), and therefore huge platforms like Amazon and Google do engage in content production, not just circulation.

[6]In a famous passage in *Capital, Vol. 1*, Marx himself suggests that the ability to plan out projects before they are undertaken is what distinguishes humans from other labouring animals. His example is of the architect who, unlike any animal, 'raises his [sic] structure in imagination before he erects it in reality' (1976: 284). This rather crude assessment of what makes humans human is undercut by more recent accounts that emphasize the interconnectedness of doing and planning, as well as of corporeal and mental activity (e.g. Sennett, 2009).

and those whose time is 'worth almost nothing' (1974: 58). Deskilling is bad because it renders workers easily replaceable, thereby keeping wages down, and excessively rewards the skills held by a select few.

DIY practitioners are not always workers. As such we should be careful not to overstretch in search of parallels, nor to focus exclusively on the economic consequences of platformization at the expense of the cultural. But what Braverman's deskilling thesis helps to illuminate is that the issue in the contemporary situation is not automation *per se*, or the hiding away of complex technological process behind user-friendly interfaces (which has some wonderful and empowering consequences). It is not about 'lost skills' or the absence of 'real work' in contemporary DIY. DIY today is not less resistant because 'everyone is doing it', or because it is 'easy' – in fact, these qualities represent the significant upsides offered by automation. Rather, the problem is with an extreme polarization of value, and with the tendency for platform capitalism to, thus far, concentrate power and wealth in the hands of a very few. Platforms offer musicians a greatly increased capacity for 'planning' and, in that sense, everyone can be DIY. But they are in no position to effectively bargain for a greater share of the economic rewards that are being collectively produced.

Optimization and social media

It has been posited that the platformization of culture is impacting not only the economic and organizational dimensions of the cultural industries, but also the form and content of the products that they produce (Duffy, Poell and Nieborg 2019, Nieborg and Poell 2018). Nieborg and Poell, for example, identify a new tendency for cultural commodities to have 'contingency' – that is, to be adaptable to the norms of platforms like Facebook and to have a particular 'modular' flexibility which allows timely accommodation and implementation of audience data (2018). Their primary examples come from journalism and video games, but others have identified comparable activity in the world of popular music, where composition and production are to some degree geared towards pleasing the algorithms of streaming services like Spotify and Apple Music (Bonini and Gandini 2019: 9, Mack 2019).

This kind of 'contingency' does not seem to be an issue in DIY, at least in terms of the music itself.[7] This section focuses on a related facet of

---

[7] DIY (and lots of other musical production) is still largely centred around slower processes, including long release times for albums. Despite some arguably 'contingent' recent releases by stars like Frank Ocean and Kanye West, I'm sceptical of the idea that (musical) creativity faces a crisis resulting from an excessively 'tech-facing' approach. The idea of writing music 'for' the key distributive technologies of the age is not new – for example, the sound and structure of twentieth century music was in part shaped 'for' radio – and always also includes consideration of a potential human audience (i.e. there is no point writing *just* for the algorithm).

platformization – namely, optimization – which does offer a particularly germane challenge for DIY music. Research on optimization to date has mostly focused on creative and cultural workers, and their 'knowledge-building and interpretative processes surrounding algorithms' (Cotter 2019: 896). Since algorithms' exact technical operations are often something of an unknown quantity, social media users are somewhat in the dark, yet they nonetheless seek to 'optimize' their activity for algorithms – activity that is sometimes labelled as 'gaming' or 'hacking' (Petre, Duffy and Hund 2019). Research has also highlighted connections between optimization and self-branding, the furtherance of socio-economic inequity (Bishop 2018), and the particular emotional burden that comes with what Taina Bucher calls the 'threat of invisibility' (2012; see also Cotter 2019 on 'the visibility game').

This threat of invisibility is real enough. But in outlining optimization as a logic (rather than just a practice), I also want to argue that this terminology of visibility is in itself a kind of concession to such logic. 'Visibility' theses tend to focus on optimization as it manifests in the relationship between users and algorithms, rather than considering optimization in the context of the political economy of cultural production – wherein producers compete *amongst each other* for the limited attention of consumers (Garnham 1990: 158). The argument I present here is that the logic of optimization drives competition between practitioners who would not otherwise be inclined to compete with one another.

I make three claims about this logic of optimization. Firstly, it reflects a re-calibration of what constitutes 'common-sense' communication, through which marketing norms are made part of everyday platform activity. Secondly, it represents a substantial knowledge transfer from platforms to users and is therefore an important means by which data is put to work by platforms. Thirdly, while optimization appears to be non-competitive, it in fact serves to accelerate internal platform markets, the end result of which is that paid promotion becomes increasingly required and normalized.

DIY practice has historically involved opting out of many of the 'norms' of music industry promotion – refusing mainstream press, refusing advertisements and demonstrating an unwillingness to frame DIY activity as a commercial venture or to engage with marketing practices in general. But the notion of what constitutes marketing, and what is simply communication, is of course complex and ever-changing. For some practitioners, the idea of self-promotion felt seedy and unethical to the extent that it was anxiety-inducing. One practitioner described attention-seeking as

> the self-aggrandising Achilles' heel of putting anything out there, the idea that you're sort of saying, 'look at what I'm doing'. Essentially you're trying to reel people in but on some level that's arguably unethical ... it's

a weird one ... like whenever I put out stuff, be it put on shows or put out cassettes ... the channels that I use are generally channels which operate around just vocal conversation so just telling people about it verbally or via the internet. (P1)

The technique used here to avoid the feeling of 'unethical' promotion is to utilize only those social channels that seem to be more 'naturally' occurring. Similarly, many practitioners only posted about new music or events once (rather than more frequently) because they didn't want to 'annoy' their online audience.

♫ 'Ambition' – **Subway Sect**

But a logic of optimization argues that there is little meaningful difference between 'getting the word out' and 'getting the word out *effectively*', insofar as they are both about avoiding 'invisibility'. This sense is exacerbated by the work of optimization being comprised of a myriad of small, relatively insubstantial decisions. This kind of optimization is described here by a practitioner reflecting on deciding the best time of day to upload a new music video to Facebook:

I think that [knowing the best time of day to post on Facebook] was just common sense to me, like I think it was maybe when we were putting up the video that we did, and it was like, we put so much effort into that video, I want as many people to see this thing as possible, so if I upload it at like midday when everyone's at work, that seems pretty pointless, so if I upload it at like 6pm, everyone's got home from work and is probably like on their phone or on their laptop or somewhere, and that just seems more obvious people are gonna be not at work then and stuff. (P27)

This practitioner was quick to stress that they placed no great value in social media attention *itself* ('I'm more interested obviously in people just hearing our music, that's the main thing, that's the reason we even have social media') and that any 'gaming' of the attention economy wasn't done cynically but reluctantly ('I kind of think it's all a bit rubbish, the fact that that kind of stuff matters [...] in the eyes of promoters').

Optimization can act as a kind of knowledge transfer between platforms and users, giving over secrets gleaned from the platforms' data mining in an attempt to shape user norms. This is often done through defaults – the default option will be the 'best' one to choose – but also takes place through suggestions. Facebook offers (often un-asked for) guidance on how to create posts that 'perform better'. Social media usage itself creates a kind of empirical knowledge of how to optimize, such as in one practitioner's observation: 'like, if you're just putting up a random post, just put up a picture with it, 'cos *obviously* people are more likely to interact with a picture than just a status' (P27). Access to metrics like Facebook and Twitter Insights then allows 'obvious' knowledge to be confirmed and reinforced.

As above, these decisions are considered 'common sense', since the information is already known and therefore allows for the enactment of a strategic, self-managerial approach to growing an audience, without falling into the more obviously problematic realm of marketing, nor having any of advertising's connotations of cynical manipulation (or *creation*) of an audience. Assuming that the new song has to be posted on social media *at some point*, and therefore at some specific time of day, it feels like a fairly small and unremarkable step to then ask: *'what time would be best?'*, where in this context the question is asked not in terms of optimal self-expression, but optimal engagement. In this way, the logic of optimization reinforces the kind of reflexive self-presentation that, as I argued in Chapter 3, constrains practitioners in ways that are harmful for claims to recognition.

An important characteristic of optimization is that it emphasizes the agency of individual online consumers. This is in part due to the common understanding of the internet as something less constraining than traditional media. Unlike broadcasting or the printed press, where we have a sense of being 'talked *at*' and 'marketed *to*', the internet seems to offer a myriad of choices that we participate in and shape in our own image, rather than passively receive (Patelis 2013: 119). The (now rather dated) metaphor of the 'information superhighway' captures this sense that the direction of travel is ultimately up to the user. Optimization is about improving one's chances of getting attention in an economy in which individual users are ultimately in charge, and can't be *told* what to do, unlike the perceived ability of more conventional marketing approaches to 'trick' consumers into changing their minds.

In this light, optimization has significant similarities with the 'nudge' theory, as developed and popularized by behavioural economists Richard Thaler and Cass Sunstein (2009). In introducing their theory, they refer to their political approach as 'libertarian paternalism' – an oxymoronic label which suggests that the liberty of 'free' choice is inviolable, but also that it can beneficially shaped by overseers. Thaler and Sunstein contextualize their theory with a vision of its future application, which imagines well-intentioned institutions 'nudging' users into the 'best' decisions, without *forcing* them to choose (2009: 4). Their influence has been substantial, and although Thaler and Sunstein caution against misuse of nudges, their approach assumes that those with the power to nudge might have a 'paternal' interest rather than any more duplicitous intent.[8]

However, the ability to effectively 'nudge' is dependent on a combination of scientific trial-and-error, having the capacity to access and analyse sufficient data, and being in a position to implement required change. Aside perhaps from governments, it is the monopolistic rulers of platform capitalism who are best placed to nudge, and controlled trials on Facebook (in collaboration with academics) have been noted and criticized by privacy advocates and researchers alike (Booth 2014, Jouhki et al. 2016). As with Frederick Taylor's 'scientific management', discussed above, the application of nudging is a specific consequence of technology put to work for capitalism. Resulting gains are likely to be skewed in favour of the existing dominant firms. Although, as highlighted above, platforms are willing, where appropriate, to share some of their findings with practitioners, thus off-loading managerial responsibility to independent cultural producers and increasing internal competition.

But the 'nudging' logic of optimization is not restricted to the very largest platforms. It is also prevalent – indeed, especially so – on the music sales platform Bandcamp, a smaller-scale sales platform with some particular affinities to independent and alternative music culture (Hesmondhalgh, Jones and Rauh 2019). For example, when listing a new release for sale on Bandcamp, a pop-up text box suggests the optimum selling price, based on their data on what prices have sold best in the past, and gives similar advice and information intended to put the artist in the best possible position to make sales. And Bandcamp's approach to presenting this kind of optimization fodder epitomizes the hands-off libertarian paternalism of the nudge. Perhaps wary of scaring off the marketing-averse independent musicians that make up their clientele, their guidance on pricing for digital downloads is preceded by this caveat:

> Please take what we're about to tell you with a grain of salt. Part of what makes Bandcamp *Bandcamp* is that you, not some corporate behemoth, set your own pricing. And that's really as it should be, since the most effective price just isn't the same for every artist, and you know your fans better than anyone. That said, we have the advantage of a metric crap-ton of data, and that data tells us a few things: […].
>
> (Bandcamp nd)

---

[8]The UK Conservative–Liberal Democrat coalition government established a 'nudge unit' in 2010 to put forward policies based on the application of Thaler and Sunstein's theory (officially named the Behavioural Insights Team, then later part-privatized and renamed Behavioural Insights Limited).

This is followed up with a guide to the best prices for '*most* artists' in order to maximally profit from downloads. The informal and self-aware tone taken by Bandcamp (e.g. the semi-ironic description of the platform as a 'corporate behemoth') is in keeping with the 'hands-off', relaxed approach that characterizes optimization, suggesting that the choice remains in the users' hands, and even anticipating some degree of distrust. This is exactly the 'paternal libertarianism' of nudge theory: we know best, and it's in your interest to listen to us, but you don't *have* to.

It is tempting to ignore the ideological content of nudges and defaults as easy to opt out of (platforms even admit as much themselves), and therefore relatively inconsequential, but this would be to ignore the key finding of this area of behavioural economics: users are particularly likely to do what they're told *when they don't think they're being told what to do*. Optimization and 'nudge' theory both suggest that the best kind of marketing is that which makes the decision feel as though it is coming from the recipient of the marketing. The following paragraph from Bandcamp's pricing guide highlights this philosophy neatly:

> While we have your attention, we would like to discourage you from doing one-penny-off pricing (e.g., $0.99, $9.99, $11.99). Though it may be an effective tactic for selling waterbeds, cell phone plans, and Angry Birds 34, when we see that sort of pricing on an artist's own website, we do not think 'gosh, this is a good deal' but rather 'what we previously thought was a person/band is actually a marketing department, and they're subtly telling us they think we're idiots.' Present a straightforward price, let fans pay more if they want, and they'll reward you.
>
> (Bandcamp nd)

The anti-marketing rhetoric offered by Bandcamp seems compliant with counter-cultural values: they understand that music is not a 'waterbed' or a 'cell phone', distancing themselves from those cynical music biz parasites who only care about the bottom-line. They also suggest an ethical dimension: be straight with your customers and, happily, you'll get a (financial) 'reward' for your judicious approach.

Of course, the 'we know you're not a product' shtick is only possible because, as outlined in the previous section, platform capitalism means their profits are relatively untethered from any given practitioners' sales. Whilst Bandcamp takes a 10 per cent cut of all downloads, they are primarily interested in the growth of the platform as a whole rather than *your* individual success or failure. And, importantly, the avoidance of marketing (i.e. the penny-off approach) is not advised because marketing is a blight on society, but because it turns out it isn't in fact the most optimal kind of marketing. In this way strategies of resistance ('I don't want my music to be commodified') are transformed into cynical gestures of alterity ('it is beneficial to *look like* I don't want my music to be commodified'). Strategies of optimization are used by platforms to 'nudge' practitioners, and in turn by practitioners to 'nudge' their audience.

Another way in which optimization retains compatibility with DIY norms is through the assurance that many strategies are not aimed at users *per se*, but at making sure practitioners' communication successfully navigates the gauntlet of algorithmic hurdles that determine what content reaches users' timelines and home pages. The logic of optimization argues that if decisions are made with the intention to 'prime' or 'game' algorithms, then it's not marketing and involves none of the pernicious psychic manipulation associated with advertising. It also obscures the competitive element of gaming the system: the market is presented not as band versus band, but as each band fighting an individual battle with the algorithms. In actuality, algorithms work as a content filtration system at the platform level rather than the individual level and, in seeking optimization, practitioners are really establishing an edge over their 'competitors' in the scene. (Of course, the difference in *intent* between algorithm 'gaming' and marketing continues to matter.)

The kinds of practices and strategies that I have considered so far in this section are not ones that DIY practitioners see as having a particularly strong political importance. As I have noted, they seem to be common sense, and they don't contradict DIY's anti-marketing stance, since optimization emphasizes user agency and the well-intentioned need to 'game' algorithms. However, I wish to argue that optimization weakens DIY critiques of commercialism by making marketing decisions unavoidable, and by encouraging practitioners towards capitalistic modes of accumulation.

One of the main ways in which optimization achieves this is by encouraging a dualistic perspective; social media activity is undertaken with one eye on what could happen next, but without committing to any grandiose, ultimate goal that would indicate an excessively self-interested approach. For example, one practitioner spoke about the process of 'setting up' a new band online – deciding where to host their music and to what extent these decisions impact their trajectory:

> You could either just put something on Soundcloud and be like, 'hey, me and my friend wrote some songs, here they are', or you could actually kind of establish ... I guess it depends on where you're aiming with the band. I've never consciously aimed at goals but I was very aware that actually we're doing a band and the idea is that we will record more songs, play some gigs and that's what we wanna do, so we may as well do this properly. Put the songs online and make it a band, give it a page and put a picture up or whatever, and invite our friends to like it, cos it feels like that's the idea, 'cos if in three months' time we want to play a gig, we want to have enough people who might be able to see that gig is happening and

know about it and come to it and that kind of thing. A lot of people said to me like, 'I think you were shrewd in the positioning of the band', and I ... that wasn't something that I ever thought about it. (P20)

Optimization doesn't mean 'wanting' or 'seeking' success – it means letting success find you by being 'in the right place at the right time', and therefore maximizing one's chances of being 'discovered'. But it also means that practitioners haven't really *done* anything, in terms of decisions relating to audience growth. As in the earlier quote, when told they are being 'shrewd', the DIY practitioner argues – with some justification – that they 'never consciously aimed at goals'.

Whilst it isn't possible to read peoples' desires from the outside, my impression is that the majority of DIY practitioners are genuinely unconcerned about 'making it'. Yet this doesn't necessarily counteract the feeling that everything could be just *slightly* better than it is now: that 490 Likes today could be turned into 500 Likes tomorrow. This balance between present and future also points towards the ways in which optimization strategies are limited by existing social norms, which might prevent practitioners from being too gung-ho in their attempts to garner attention. Strategies which veer too closely towards promotion, for example, are met with disapproval:

Like I hate when people on Twitter ... it's bad Twitter etiquette if they have an event and they tweet individual people like 'this is happening', 'this is happening', for the same event, it happens all the time. It's such bad etiquette, and it spreads the word about the event but I don't ever see it as being that successful really, I think it makes you look a bit desperate. (P16)

Negative feelings towards excessive posting and similar strategies are a reflection of online 'politeness norms' (Baym 1993: 157), and also consequence of DIY's own ethical norms. Decisions about how to present one's self online require the imagining of different audiences to be prioritized (e.g. 'if a fan/peer/label boss saw our Page, what would they think?'). Attempts to address this 'context collapse' (Marwick and boyd 2010) is in part what keeps optimization in check: the scene's 'moral compass' figures discussed in Chapter 4 might be watching and might disapprove.

But for the most part optimization doesn't really contradict 'DIY', at least in the sense of doing it yourself. As I argued in the previous section, DIY is in some senses highly compatible with platform capitalism, and the term 'DIY' doesn't begin to critique the internalized self-management which help platforms to thrive. If anything, it serves to legitimate them, with the emphasis on 'yourself' positing a self-sufficient managerial strategy as preferable to outsourcing such work. Optimization is specifically well-suited to DIY, insofar as it allows for the potential of growth without taking the more explicitly censured routes of advertising and third-party assistance. This demonstrates the extent to which, as Klein et al. note, digitization complicates notions of 'selling out', as marketing techniques are increasingly embedded within everyday practices (2017).

♫ 'Work it' – Marie Davidson

Scene-specific anti-marketing norms are further mitigated by the fact that optimization feels like a personal choice. It is difficult to find the *political* issue with optimization. What is the harm, exactly, with thinking about the best time to post a song? My interviewees across the board tended to see these decisions as primarily a reflection of personal comfort levels, relating to different emotional thresholds for seeking or avoiding public attention. Practitioners do form moral views on the basis of how much promotional work is visible (bands might be said to be 'trying too hard', or being 'a bit *pro*'), but in general these will be kept close to the chest.

I do think that there is something pernicious to optimization, which highlights a fundamental tension between DIY ethics and 'platformized' DIY activity. Clearly, though, the solution would not be to avoid communicating altogether. And neither would it be particularly beneficial to deliberately de-optimize – that is, posting at silly times or 'hiding' music away – although it might bring a sense of autonomy and some subjective relief (as in the anonymity discussed in Chapter 4). It is also important to note that many algorithms currently have a bias against LGBTQ content that is relevant for the many DIY scenes with a high proportion of queer practitioners – platforms' haste to mark such material as 'adult' has significant consequences for visibility and representation (Bishop 2018: 71).

The impossibility of 'resolving' the problem helps us understand its key function: optimization plays an ideological role in sustaining participation in a competitive platform economy. In a recent article outlining the continued need for media studies to engage with the concept of ideology, Downey and Toynbee distinguish between 'persuasion', that is, 'encouraging another agent to see the world in a particular way' and the 'economic and political power' that 'employs material resources as either sanctions or inducements in an attempt to get agents to behave in particular ways' (2016: 1266). Platforms' material resources, in terms of communicative efficacy and reach, are of course an inducement to participation; optimization, though, forms part of the persuasive effort.

Optimization masks our individual contributions to internal platform competition in several ways: by positioning activity as 'against' algorithms rather than as direct competition with other users; by 'gifting' data-knowledge to users in ways that present as common-sensical rather than commercial; and by emphasizing individual consumer freedom to choose rather than acknowledging the actual, *collective* role of producers in creating and shaping consumption practices. In this way, optimization allows for a disavowal of agency and a voluntary submission to platform logics.

The erosion of critique?

This chapter has outlined some significant, perhaps existential threats to DIY music ethics, which relate closely to the economic and cultural consequences of platformization. DIY emerged as critique of what was perceived to be a largely Fordist (or Ford-*ish*) model of industrial cultural production, but much of contemporary DIY practice – which still draws on those critiques – is less effective at countering the new economic arrangements that underpin the 'platformization' of culture. The new discrepancies between DIY activity and platformized culture are not fault lines but grey areas, wherein self-sufficient approaches to musicking slowly drift – sometimes unwillingly, sometimes unwittingly – towards individualist, entrepreneurial strategies of self-marketing and self-branding. The challenge, then, is significant. But the pessimism that characterizes this conclusion is hopefully offset by the final chapter, which offers some suggestions as to how DIY might, in light of these challenges, continue to orient itself towards social justice.

8

The plan ...

Potential futures for DIY music, new media and social justice

A key aim of this book has been to show that, even though social media has assisted in creating a situation in which many music practitioners are 'DIY by default', DIY music scenes do remain ethically and practically distinct in some important ways. However, that distinctiveness is placed under significant architectural pressure by specific platforms, in terms of the way that metrics and algorithmic news feeds encourage and reward individualistic competition. It is placed under a broader pressure by the cultural and economic logic of platform capitalism, which calls into question the political meaning of independence and 'doing it yourself'. In the last, short chapter of this book, I make some suggestions regarding how this distinctiveness, or at least the valuable aspects of it, might be protected. Although there might be large-scale solutions to the problems posed by platforms – with the idea of breaking up or even nationalizing Facebook and Google currently part of mainstream-left political agendas in the UK and the United States – this chapter focuses on approaches that might feasibly be taken up by DIY practitioners themselves. The suggestions here do not constitute anything like a comprehensive manifesto, but are intended to contribute to and further a debate that is already underway within many DIY scenes and alternative music communities across the world.

In Chapter 2, when outlining a history of DIY music, I suggested that there were some tensions that ought to be considered highly characteristic of DIY, which arise primarily as a consequence of DIY's ambivalent relationship with popular music. One of these tensions is between *resourcefulness* and *refusal* – that is, between finding political value in doing more, or in doing less. That tension serves as a structuring device in this chapter, which considers firstly how DIY practitioners might resourcefully build alternatives outside of the current platform economy, and secondly how they might refuse to comply with the logics of the dominant social and musical platforms they use today.

DIY music cultures have historically valued alternative distribution networks, and the building and maintaining of spaces 'outside' of the commercial music industries. The potential for building alternative music distribution networks online may feel lost, but I argue that there are possibilities that remain under-explored, and a strategy of resourcefulness could start here. Building an entire 'platform' might be beyond the scope of DIY. But there are a number of existing sites and tools that highlight this potential for alternative distribution, from the commonly used to the experimental. DIY, at least in the indie-punk tradition, is hindered in this regard by its scepticism towards open-source and creative commons approaches. I suggest that this could and should change.

A strategy of refusal might hinge on challenging the elisions implied by social media between work and leisure as well as between amateur and professional music cultures. Understanding DIY culture as a non-work activity would allow us to question the aspirational practices of optimization that characterize 'good' behaviour on social media. It would also mean challenging the apparent inevitability of professionalized temporal structures – for example, the recording and promotion schedule – that are increasingly dominant in DIY activity. Ultimately, some combination of resourcefulness and refusal might be most appropriate (and most feasible) for maintaining and strengthening DIY's connection to issues of social justice.

Building alternatives

The primary argument against DIY resourcefulness, historically, has been that it plays into the hands of its antagonists – that is, the major labels which periodically swoop down to steal away the most marketable aspects of DIY music. Crudely put, the argument is that the harder DIY practitioners work to create meaning, the quicker this surplus value will be siphoned off. But the predominant role of platforms in DIY music culture today substantially alters this proposition. The kind of co-option described above is centred on a particular moment of exchange, whereas the interactions between DIY music and platforms are not 'momentary' in the same way. They are constant. Platforms do not come along and 'take' only the most valuable aspects of DIY, but instead benefit from DIY activity at all scales.

One resourceful solution to this problem of operating 'alongside' platforms might be to construct alternatives to monopolistic social media platforms, in doing so regaining control over distribution. DIY music has scarcely ever been about aiming to overturn entire systems of production, and therefore creating 'a new Facebook' is almost certainly beyond its scope. However, it has managed at various times to construct a space where market dynamics are less pervasive. DIY might find similar success today by attempting to ensure that Facebook and YouTube are less dominant online in the areas where it seems most pertinent – making sure that these platforms aren't the *only* places that bands are manifest online, that the music and the other important artefacts of the scene are hosted elsewhere (especially important given platforms' increasingly all-encompassing terms of service) and that (cultural as well as economic) value generated by DIY activity is kept at arm's length from monopolistic corporations.

Such alternatives might be based on FLOSS (Free/Libre Open-Source Software) principles. In particular, the 'free/libre' part is critical, since this identifies not only that the code is free to use, but also that it must remain free and open if it is to be re-used or adapted. It is the free/libre aspect (and the related notion of 'copyleft') that distances open-source projects from the most nefarious kinds of 'crowdsourcing', where the knowledge, content or data generated by the collective is then brought back into private ownership and capitalized upon (see critique of Tapscott's 'wikinomics' in Taylor 2014: 22–3). Google's mobile operating system Android, for example, is built upon open-source technology, but since there is no 'free/libre' component to the licensing, Google are able to create their own 'forks' (i.e. modified versions) based on the publicly shared code which they then make proprietary. The UK government's Open Data initiative similarly allows for both commercial and non-commercial use; its claim to be 'opening up government' is in practice allowing private enterprise free access to a growing archive of public data. Any new DIY platform-building project would need to be careful about where and when 'free' data means a libertarian freedom to privatize and capitalize on the work of others.

Cooperatively owned peer-production online may seem to have had its day. The more optimistic proclamations of its potential in the early 2000s failed to be borne out by evidence of such projects flourishing – with Wikipedia being the main exception (Benkler 2006, Bruns 2008b). But cooperatives do have a long tradition in DIY cultures, and the notion of 'streaming co-operatives' seems to be graining traction, as a broad range of musicians and industry actors seek alternatives to the low pay-outs offered by dominant platforms today. One example is Resonate, a music streaming platform cooperative with the tagline 'this is democratic capitalism' (Resonate n.d.). The business model as outlined on their website appears to work out better for both listener and musician, but their admission that 'we have no way of knowing exactly how this will actually break down' suggests that their pricing strategy has yet to be proved in actuality (Resonate 2015).

The difficulty facing streaming cooperatives is that same as that identified by Rosa Luxemburg over a century ago: 'Small units of socialized production within capitalist exchange' are to a large extent bound by the pressures of competing within the market (Luxemburg 2006: 47). That is to say, cooperatives tend to be hamstrung by their need to offer viable alternative to both producers *and* consumers simultaneously, while working in an economic system where consumer benefits tend to be delivered through the exploitation of producers.

In considering how an alternative to monopolistic platform capitalism might be constructed, it is worth considering why viable alternatives seem absent thus far (or at least, why such an alternative hasn't been successful within DIY circles). One practitioner here is attempting to answer that question:

> I guess if there was a kind of non-profit, open source video streaming service, we would all use it. And I guess it's about what people put their energy into creating, and what's seen as having value. Cos I guess Wharf Chambers was set up because sufficient people wanted to set up a workers' co-op bar that met these needs so they worked towards it and found a venue and set it up and now it's running successfully. And I guess maybe that's something that's seen to have more value? Or maybe more glamour? [...] Whereas whilst there's also arguably a need for open source, non-profit software to share music, I dunno ... that's not gonna get you dates. That's not gonna get you a really good social life, is it? Or, you're not gonna get a wage from that. So that's why people in our community aren't working towards doing that. And it could also be about the skills required I guess. Not saying that it's not a skilled profession working in a bar, but it's very different to the specific computer skills you would need to try and create an alternative to YouTube. (P9)

There are some important points raised here. Firstly, that technological work of this kind requires a specific and advanced skillset. There are issues of unequal participation here, not just in terms of allowing practitioners to play an active part in building and maintaining a platform, but also in terms of developing the front-end usability that would make it a feasible option for those looking to distribute their music. There *are* existing open-source, non-profit alternatives to YouTube (LBRY, MediaGoblin), but they are sufficiently unintuitive and cumbersome as to not constitute an alternative in any viable sense. Also, many of these alternative sites emerged as responses to a perceived crackdown on 'free speech' on existing platforms; as such, they are often populated by people wishing to share and create material that is hateful and discriminatory.

Secondly, the above practitioner identifies the role of computer programmer as uncool, anti-social and financially unrewarding. This lack of interest in coding-as-resistance within DIY highlights that any project to develop platform alternatives would need to be able to fully articulate its beneficial consequences for social justice. Here a solution from *within* DIY might involve using the scene's significant capacity to create symbolic meaning and bestow countercultural capital in order to make open-source coding feel slightly more exciting and more in keeping with DIY's own ethics. This might be achieved by emphasizing skill-sharing (like the existing DIY skill-sharing workshops), redressing the tech sector's gender imbalance, understanding the harmful biases that are reflected and reinforced in platforms' algorithms (Noble 2018) and the ways in which platforms' emphasis on free speech and self-expression creates a problematic false moral equivalency. In order to present itself to DIY practitioners as 'cool' (i.e. worth doing), any resourceful alternative would need to demonstrate the cost of using existing social media at the individual, collective, societal and global level, and show the need for change as *urgent*.

If entire platforms are too difficult to build, or require more commitment than could be expected, then there are opportunities to resist on a smaller scale and to build on the work of others, for example, electronic musician Mat Dryhurst's platform Saga, which allows users 'full control over how your videos behave in each different place they're embedded online' (Dryhurst 2015). So, unlike YouTube and SoundCloud content that can be embedded within any website with or without your consent, using Saga means that if you don't like the context in which your video is shown, you can make it behave differently on that particular website. In highlighting this capacity, Saga re-presents the apparent neutrality of 'sharing': being disinterested in how and where we allow our content to be used makes us complicit. Saga allows users to respond with a firm 'no', rather than the tacit 'okay' we give when using the usual platforms. Of course, Saga's functionality currently leaves something to be desired: 'A sacrifice you make in using [Saga] today is that it is young. It doesn't yet resolve prettily on Facebook. It can be very buggy when viewed on mobile phones. It doesn't *just work* – yet. You may have to play with it a little to get what you desire out of it. When something "just works," ask yourself – for whom?' (Dryhurst 2015).

However, as with the possibility of countering coding's 'uncool-ness' by highlighting its political potential, there is an opportunity here to redefine the *materiality* that DIY has historically emphasized, and which is now seen as absent. The need to get one's hands dirty with code (rather than using something that will 'just work') might to be seen as politically analogous to the back-of-the-van materiality of touring in the 1970s and 1980s.

Any alternative would also need to acknowledge the extent to which platforms' current functionality provides significant benefits for practitioners:

> I guess for me, in the same way that there's no such thing as ethical consumerism, we're stuck in this really shit capitalist society, where anything you make can potentially be capitalized on, and a lot of the free tools that we have at our disposal, things like YouTube or Bandcamp, the trade-off is that someone's profiting from them. And I guess for me that's a sacrifice I'm willing to make, but it's not ideal. And I think it's still ... for me the fact we have things like laptops where you can easily set up a webcam and recording yourself playing and put in on YouTube, or record stuff on your computer in your bedroom and put it on Bandcamp and sell it to your friends, is ultimately really positive. And I think it's good to be aware of who else might be profiting from it, but I can't see a way to distance yourself from it without kind of, cutting your nose off to spite your face. (P9)

This practitioner highlights something that has been a theme all the way through this book: there are really positive dimensions of social media that we might not want to lose. Platforms offer us instantaneous mass communication, widespread access to creative audio-visual tools, and the capacity for collaboration and the exchange of ideas. These are the features that were at the heart of the early optimistic claims about Web 2.0 and social media's radical democratic potential, and the challenge is to build something that doesn't involve losing these beneficial aspects. This also emphasizes that there is no 'going back' from platforms: whilst local music forums functioned as a decentralized home for DIY activity in the early 2000s, they would by today's standards be too cumbersome and inflexible to compete, and would involve an unnecessary repetition of effort. Platforms' great achievement is that they are built only once and provide space for all, and thereby permit, in Paul Virno's terms, 'a communality of generalized intellect without material equality' (2004: 18). Any new alternative must take this 'communality' as its starting point.

The other difficulty raised by the interview quotes above is how to leave these existing platforms when the cost of leaving is currently so high. Langlois argues that 'for-profit social media are just too much a part of our lives for us to do without, and too complicated and expensive for us to construct alternatives'. For many of us, our friendships and relationships are so embedded within the platforms that boycotting or 'giving up' platforms would mean 'missing out, quite literally, on our lives' (2013: 52). However, this problem contains its own solution, in that the truly irreplaceable part of social media is *the social* – that is, ourselves, and the connections between us.

However, there's also no guarantee that open-source platforms would be radically different in terms of the features they offer and the kind of subjectivity they are built to enable. The most successful attempts at open-source social networks tend to stick closely to the features found on successful proprietary platforms, and which therefore are understood to be (explicitly or implicitly) 'demanded' by potential users. As such, any negative consequences for recognition arising from the architecture of the platform are likely to be recreated in an open-source equivalent. Hui and Halpin ask provocatively: 'If Facebook, as the predominant example of a centralized digital social networking platform, is to be considered the apex of the industrialization of social relationships, can users escape their reduction to social atoms by simply decentralizing Facebook?' (2013: 107).

Hui and Halpin argue that the specific subjectivity formed by social media is primarily a consequence of the network itself, and its manifestation of the social as an accumulation of connections between nodes, and therefore that public ownership wouldn't necessarily bring about substantial change. Berry argues that networks inherently 'encourage users to think of themselves as a set of partial objects, fragmented "dividuals," or loosely connected properties, collected as a time-series of data-points and subject to intervention and control' (2013: 44).

What might a FLOSS platform do for claims of recognition, in the context of DIY music, that existing dominant platforms can't (or won't)? Especially given the fact that we know there is much that it probably *can't* do, in terms of user friendliness and size of its user base. Critically evaluating networks is of value, but much of the theory dealing with these issues is operating at a high level of abstraction, and therefore some of the calls for 'counterprotological code' (Galloway and Thacker 2007: 100) seem a long way from anything that might feasibly be acted upon.

But things can change fast. Even relatively recently the free exchange of information online, unfettered by copyright restriction, seemed to many like an inevitability. Peer-to-peer file sharing was near-enough a banal normality, before existing media giants stepped in, using their existing wealth and power to lobby for the re-affirmation and extension of copyright law (Vaidhyanathan 2004, Prior 2015). Diane Gurman highlights the importance of 'framing' in this context, where moral codes are applied in ways that suited vested interests, noting that 'news stories since the passage of the CTEA [the United States's 1998 Copyright Term Extension Act] tend to brand any unauthorized use of copyrighted material as "piracy," seldom mentioning the public's right to access information, and forgetting that copyright law also includes legitimate exceptions, such as fair use' (Gurman 2009). The 'framing' of social media has been equally successful, even in the ways that key terms like 'platform' (Gillespie 2010, 2017) and 'sharing' (John 2013, 2017) piggyback on the communal connotations of peer-production, without drawing attention to the question of what we are sharing with whom and why. The challenge in building an alternative is to highlight the political and cultural implications of 'sharing' as a metaphor and create new spaces in which to practise *actual sharing*.

Using existing platforms differently

So, new strategies of resourcefulness might include building alternative distribution networks, bringing forms of cooperative and collective ownership that are already present offline onto social media, and taking control of 'code' in order to emphasize the materiality of resistance online. However, there are elements of refusal that might still prove valuable in countering some of the difficulties discussed in the chapters above, and in offering more 'everyday' tools for carrying out cultural resistance.

As I have identified, existing strategies of refusal are often undermined on social media because platforms find value through circulation and data, rather than the specific activities which within DIY are demarcated as 'productive'. It's also unclear when data is collected, and therefore when and how one is opting out, or what the most effective strategies might be. Additionally, the sheer scale and reach of platforms seem to undermine efforts to refuse to participate, casting them as irrelevant and ineffective. Therefore, new strategies of refusal ought to focus not on the economic consequences of boycotting social media or disrupting from within, which are likely to be minimal, but on rejecting the new forms of subjectivity created by social media, and on the positive consequences for recognition of the self and others that might result.

So, firstly, a new politics of refusal would acknowledge that the sense of haste generated by social media is often something that primarily serves the interest of the platforms, reinforcing the specific temporalities of immediacy and newness – what Kaun and Stiernstedt call 'Facebook time' (2014) – that support communicative capitalism. Additionally, we might conclude that to refuse to rush, and to give one's self more time to contemplate, might have beneficial effects in terms of self-realization. One practitioner used this kind of argument to explain why their band had only a limited social media presence:

> I think people tend to, broadly speaking, make their best creative work when they have time to just, like, sit. Not even thinking about it or dedicating to it necessarily, just sit and be with themselves: the eureka moment is when someone, an intelligent thoughtful person who is dedicated to their craft, has that small pocket of time where they're between inactivity and thinking about what they're doing, and that's when the inspiration comes. And it becomes more and more difficult to do that when you feel this constant need to prove that you 'have done' or that you 'are doing'. Which, you know, is a classic trap that people fall into on Twitter or Facebook. (P25)

The issue here is not with rushing per se, but in rushing towards something that is perceptibly an 'outcome'. As with the 'Slow University' movement

(Treanor 2008, Mendick 2013), the political nature of the problem is in who determines what qualifies as an outcome, and the way the resulting pressures shape our activity. The challenge is to resist this outcome-driven approach or to re-define outcomes on one's own terms – something that DIY practitioners are sometimes able to do. One Leeds-based DIY musician was recounting to me how much they had enjoyed 'practicing' with some friends in an as-yet-unnamed band, who were playing together but had no plans to perform in front of an audience:

Interviewer: I wonder if at some point that becomes not practicing but something different? if you've got no shows on the horizon, it's not practice is it, it's something else.

P17: Yeah exactly, it's a case of just playing together and it's also a really nice structured way to hang out with people and get to know people, cos you're all, at least when you're presenting ideas, you're in a very vulnerable position I think, and I think that's a really nice way to bond with people. It's very supportive and collaborative and I realized that yeah, I'd never really felt that way before about being in a band.

My question was perhaps a leading one. But seeing music as 'a really nice structured way to hang out' points towards the social aspects of DIY and towards an experiencing of these moments in the present (rather than preparing to re-present them in the future). This might also involve a re-focusing on music-making as a specific and unique practice, with benefits to the self and to relationships that aren't offered by the new practices of self-promoting on social media. One practitioner felt conscious of the way the internet's compulsive pull might detract from music-making time:

So, I could beat my head against the computer constantly every night in the studio and be like, there's an endless list of blogs [to contact], I'll just do this blog ... but then there's the guitar there, like well, but you're a musician as well, so why don't you do more of that. So, the internet can be a gift and a curse, perhaps that's what other people have said to you, I don't know. (P16)

Several practitioners mentioned being grateful or relieved that a lack of phone signal or Wi-Fi connection in rehearsal or recording spaces – in some cases a deliberate choice – meant they were able to focus on the enjoyable and rewarding activity of music-making. Again, existing strengths of DIY can be built upon here, particularly the scene's awareness of mental health issues (and how to offer support), and its ability to create and use language to discuss these problems, where existing discourse seems inappropriate. There is a related need, as Mark Fisher identifies, to identify and understand the extent to which mental health issues might have structural causes (2009: 21–5). In the context of DIY and social media then, often-expressed feelings like 'FOMO' (fear of missing out) need to be understood as exacerbated by existing social structures (and, as a result, *collectively* experienced), rather than as a personal failing.

Secondly (and relatedly), a new refusal would involve developing an awareness of the ways in which social media can shape subjectivity. This would include a close examination of practices related to self-branding, and an understanding of how feelings of pride and ambition might, in certain circumstances, work to support platform capitalism and offer little reward beyond an increased reliance on the platform as a source of validation (as explored in Chapter 5). This is more or less the definition of 'subjectivation' offered by Ganaele Langlois:

> Subjectivation [...] means fitting within the logic of social media platforms through continuous status updates, accepting recommendations, clicking on links, etc., overall, through continuous use of the platform. Such good behaviours can be rewarded: if I invite other people to use a social media platform, then I can get bigger storage for my account or credit for purchases, and other perks. Subjectivation takes place when we are invited and encouraged to adopt specific modes of usership – ways of expressing ourselves, ways of valuing the informational logic of the platform and its recommendation system, and ways of relating to others. One of the biggest perks of being a 'good' user is to be recognized and seen by the rest of the network: the more I contribute on Facebook and interact with peers and accept lack of control over my own data, the more prominently my contributions will be featured.
>
> (Langlois 2013: 56)

I have already shown that DIY practitioners can and do sometimes resist the encouraged 'specific modes of usership' – for example, Facebook Likes can take on a moral meaning that is specific to DIY (see Chapter 6). A greater awareness of the distance, where it exists, between the 'encouraged' mode and 'resistant' mode of usership is necessary if something like a 'moral economy' (Kennedy 2016: 111–13) of DIY is to be able to continue to inspire interventions.

A strategy of refusal might interrogate the new divisions, or lack of division, between work and leisure, and how changing conceptions of work and leisure time might impact on the political status of DIY music. Work and leisure have long constituted a problematic dichotomy under capitalism, with the demands of work leaving leisure often experienced as the compulsion to enjoy one's self (sometimes to excess), or the imperative to relax (in order to recuperate for work). Literature on 'prosumption' points towards the ongoing erosion of work-leisure distinctions (Ritzer and Jurgenson 2010), although it is the autonomist Marxist focus on 'immaterial labour' that makes a greater effort to capture the resultant impact on subjectivity. In their claim that immaterial labour is now so all-encompassing that 'living and producing tend to be indistinguishable', Hardt and Negri undoubtedly overstate the case (2005: 148). However, my research does suggest that the future-oriented perspective which characterizes social media is a means by which DIY practice is 'career-ified' – even if practitioners aren't expecting (or wanting) a career in music, they nonetheless follow certain paths and routines of growth, with an emphasis on being risk-averse and maintaining a stable and branded identity.

A recent report on mental health within the music industry concluded that, in the current economy, 'music making is therapeutic, but making a career out of music is destructive' (Gross and Musgrave 2016: 12). There is a danger that, at its worst, DIY music involves all the self-managerial stress of a career in the cultural industries, with many of the same pressures of gaining and keeping attention, and without the financial reward. Given that the odds of getting 'work' from DIY are slim and that the conditions of any resulting work are likely to be precarious and unhealthy, a new refusal might more explicitly reject the 'rewards' on offer, in favour of enjoying the more immediate pleasures of sociality and an unfettered self-expression. In short, a new refusal might ask what DIY music would look like if we acknowledged that there was nothing to lose.

♫ 'We are real' – Silver Jews

This in turn might lead us to reflect on what kinds of transparency and opaqueness we employ on social media. Much of DIY music's social media activity is not about 'sharing' but 'hiding' – keeping something as one's own until the point at which there is value to be gained by making it public. Existing strategies of optimized attention-seeking are premised on this calculated retention of information: the building of hype when someone is about to 'drop' new music; stockpiling photos in order to post one per day; in this way exacerbating the future-oriented and risk-averse character of DIY music today. But 'sharing' also has the potential to exacerbate problematic dimensions of communicative capitalism – that is, the compulsion to be reflexively 'seen doing', and the sense that nothing has really 'happened' unless it is witnessed by others online. I am not calling for complete transparency as a means of creating collective feeling, since there is no evidence to suggest this would be the result. Rather, I am arguing for an assessment, based on the aims of social justice, of the ways that we over-share (in terms of surveillance and data collection) and the ways that we under-share (withholding information in order to maximize its value).

The final issue which might also relate to a strategy of refusal is an increased awareness of the environmental consequences of social media and data usage. Nick Srnicek suggests that 'data is quickly becoming the 21st-century version of oil' (2017b). However, it looks more likely that the twenty-first-century version of oil will still be oil, and that energy companies will continue to exploit depleting fossil fuel resources at high cost for the world's human and non-human inhabitants, in part in order to sustain our internet dependencies.

DIY indie-punk has, both historically and presently, tended to be fairly unengaged with environmental issues. It has been sceptical towards the tech-enabled utopian visions of dance music, but it has been even more sceptical of the romantic anti-technologist tendencies of genres like folk and country – tending to see positive change coming through ambivalent engagement with technology, rather than the avoidance of it. Additionally, those with progressive politics, particularly concerning issues of identity and recognition, are perhaps rightly sceptical of the prejudicial norms that might be elided in the 'back-to-nature' rhetoric of much environmental campaigning. However, a strategy of refusal might benefit from understanding environmentalism as part of a critique of global capitalism, and developing an awareness of how even the most democratic online platform might nonetheless, as a consequence of its contribution to climate change, be counter to goals of social justice (Devine 2019).

DIY and social media: A new ambivalence

DIY has historically been defined by ambivalence towards the popular music industries. It has found things to value and emulate in the best of popular music – its exuberance, its communicative speed, its political resonances, its communality and much more – whilst also decrying the industries' broad subordination of artistic expression to commercial profitability. DIY practitioners have sought to build and sustain an ethically preferable strain of popular music premised on reducing the distance between the production of popular musical culture, and the real needs and desires of people. This ambivalent relationship has persisted for close to fifty years.

DIY music is now also in an ambivalent relationship with the ICT industries – another culturally prominent communicative form that seems to carry democratizing potential, but which is also imbricated in injustice and discrimination on a global scale. This ambivalent relationship is in a much earlier stage. And while previously it was DIY that 'took' from popular music, today it is social media that seems to 'take' from DIY and from the principles of alternative culture more generally. Old values of independence and participation are no longer meaningfully resistant in and of themselves; they are fundamental to the operation of still-emerging forms of techno-capitalism.

♪ 'What do we do now?' – **The Just Joans**

This changing context has raised pressing questions regarding what aspects of DIY culture to protect, and what to critique. The need for music cultures to 'resist' many harmful norms is as strong today as ever, in terms of both maldistribution and misrecognition (the two axes on which Nancy Fraser builds her concept of social justice). But it may be that the values and aspirations that were previously articulated to independence – fair remuneration, self-expression, diverse participation, creative freedom and more – are best actualized now through some other means. The aim must not be to retain DIY as a musical tradition simply for its own sake, but rather to strive to make the strongest possible connection between our musical cultures and the distant, admirable goal of social justice.

REFERENCES

Abercrombie, N. and Longhurst, B. (1998) *Audiences: A Sociological Theory of Performance and Imagination.* London: Sage.

Adorno, T. W. (2005) 'Scientific Experiences of a European Scholar in America'. In *Critical Models: Interventions and Catchwords*, edited by Pickford, H. W. New York: Columbia University Press, 215–44.

Allen, R. (1996) *Delta 5: Interview by Mike Appelstein* [online] available from https://www.furious.com/perfect/delta5.html [18 February 2020].

Anand, N. (2005) 'Charting the Music Business: Billboard Magazine and the Development of the Commercial Music Field'. In *The Business of Culture: Emerging Perspectives on Entertainment, Media, and Other Industries*, edited by Lampel, J., Shamsie, J. and Lant, T. K. Mahwah, NJ: Lawrence Erlbaum, 139–54.

Andersen, M. and Jenkins, M. (2001) *Dance of Days: Two Decade's of Punk in the Nation's Capital.* New York City: Soft Skull Press.

Andrejevic, M. (2004) 'The Work of Watching One Another: Lateral Surveillance, Risk, and Governance'. *Surveillance & Society* [online] 2 (4): 479–97.

Andrejevic, M. (2007) 'Surveillance in the Digital Enclosure'. *The Communication Review* 10 (4): 295–317.

Arola, K. L. (2010) 'The Design of Web 2.0: The Rise of the Template, The Fall of Design'. *Computers and Composition* 27 (1): 4–14.

Arnold, G. (1993) *Route 666: On the Road to Nirvana.* New York: St Martin's Press.

Arvidsson, A. (2008) 'The Ethical Economy of Customer Coproduction'. *Journal of Macromarketing* 28 (4): 326–38.

Azerrad, M. (2001) *Our Band Could Be Your Life: Scenes from the American Indie Underground 1981–1991.* London: Little, Brown.

Bandcamp (nd) *What Pricing Performs Best?* [online] available from https://bandcamp.com/help/selling#pricing_performance [25 August 2017].

Banks, M. (2007) *The Politics of Cultural Work.* Basingstoke: Palgrave Macmillan.

Banks, M. (2010) 'Autonomy Guaranteed? Cultural Work and the "Art–Commerce Relation"'. *Journal for Cultural Research* 14 (3): 251–69.

Bannister, M. (2006a) '"Loaded": Indie Guitar Rock, Canonism, White Masculinities'. *Popular Music* 25 (1): 77–95.

Bannister, M. (2006b) *White Boys, White Noise: Masculinities and 1980s Indie Guitar Rock.* Aldershot: Ashgate.

Barney, D. (2010) '"Excuse Us If We Don't Give a Fuck": The (Anti-) Political Career of Participation', *Jeunesse: Young People, Texts, Cultures*, 2(2): 138–46.

Bauwens, M. (2005) 'The Political Economy of Peer Production'. *CTheory* [online] 1000 Days of Theory (26), available from <www.ctheory.net/articles.aspx?id=499>.

Baym, N. K. (1993) 'Interpreting Soap Operas and Creating Community: Inside a Computer-Mediated Fan Culture'. *Journal of Folklore Research* 30 (2/3): 143–76.

Baym, N. K. (2007) 'The New Shape of Online Community: The Example of Swedish Independent Music Fandom'. *First Monday* 12 (8).

Baym, N. K. (2013) 'Data Not Seen: The Uses and Shortcomings of Social Media Metrics'. *First Monday* (10).

Baym, N. K. (2015) *Personal Connections in the Digital Age*, 2nd edn. Cambridge: Polity.

Baym, N. K. (2018) *Playing to the Crowd: Musicians, Audiences, and the Intimate Work of Connection*. New York: New York University Press.

Becker, H. S. (1982) *Art Worlds*. London: University of California Press.

Benjamin, W. (1970) 'The Author as Producer'. *New Left Review* 1 (62): 1–9.

Benkler, Y. (2006) *The Wealth of Networks: How Social Production Transforms Markets and Freedom*. London: Yale University Press.

Berry, D. M. (2013) 'Against Remediation,' In *Unlike Us Reader*, edited by Lovink, G. and Rasch, M. Amsterdam: Institute of Network Cultures, 31–49.

Bess, G. (2015) *Alternatives to Alternatives: The Black Grrrls Riot Ignored* [online]. *Vice*, available from https://broadly.vice.com/en_us/article/alternatives-to-alternatives-the-black-grrrls-riot-ignored [18 September 2016].

Biafra, J. (2000) *Become the Media*. [CD album]. Alternative Tentacles, VIRUS260CD.

Bikini Kill (1991) *Revolution Girl Style Now!* [Cassette album]. Self-released.

Birch, I. (1979) 'Rough Trade Records: The Humane Sell'. *Melody Maker*, 10 February.

Bishop, S. (2018) 'Anxiety, Panic and Self-Optimization: Inequalities and the YouTube Algorithm'. *Convergence: The International Journal of Research into New Media Technologies* 24 (1): 69–84.

Blauner, R. (1964). *Alienation and Freedom: The Factory Worker and His Industry*. Chicago: The University of Chicago Press.

Bode, K. (2019) 'Streaming Exclusives Could Double Piracy Rates, Study Warns'. [1 October 2019] available from https://www.vice.com/en_us/article/xwe4x7/streaming-exclusives-could-double-piracy-rates-study-warns [15 February 2020].

Boehringer, J. (2015) 'Liberation through a Lack of Interest: The No-Audience Underground'. [17 August 2015] available from https://multiplesystemsofevents.wordpress.com/liberation-through-a-lack-of-interest-the-no-audience-underground/ [3 January 2020].

Bolin, G. and Andersson Schwarz, J. (2015) 'Heuristics of the Algorithm: Big Data, User Interpretation and Institutional Translation'. *Big Data & Society* 2 (2): 1–12.

Boltanski, L. and Chiapello, É. (2005) *The New Spirit of Capitalism*. London: Verso.

Bonini, T. and Gandini, A. (2019) '"First Week Is Editorial, Second Week Is Algorithmic": Platform Gatekeepers and the Platformization of Music Curation'. *Social Media + Society* 5 (4): 1–11.

Booth, R. (2014) *Facebook Reveals News Feed Experiment to Control Emotions* [online] available from https://www.theguardian.com/technology/2014/jun/29/facebook-users-emotions-news-feeds [7 September 2017].

Born, G. (1993) 'Against Negation, for a Politics of Cultural Production: Adorno, Aesthetics, the Social'. *Screen* 34 (3): 223–42.
Bourdieu, P. (1984) *Distinction: A Social Critique of the Judgement of Taste*. Cambridge, MA: Harvard University Press.
Bourdieu, P. (1995) *The Rules of Art: Genesis and Structure of the Literary Field*. Stanford, CA: Stanford University Press.
boyd, d. (2011) 'Social Network Sites as Networked Publics: Affordances, Dynamics, and Implications'. In *A Networked Self: Identity, Community, and Culture on Social Network Sites*, edited by Papacharissi, Z. New York: Routledge, 39–58.
boyd, d. (2013) *It's Complicated: The Social Lives of Networked Teens*. London: Yale University Press.
boyd, d. (2017) 'Hacking the Attention Economy'. Data & Society [online] available from https://points.datasociety.net/hacking-the-attention-economy-9fa1daca7a37.
Braverman, H. (1974) *Labor and Monopoly Capital: The Degradation of Work in the Twentieth Century*. London: Monthly Review Press.
Brown, W. (1995) *States of Injury: Power and Freedom in Late Modernity*. Princeton: Princeton University Press.
Bruns, A. (2007) *Produsage: Towards a Broader Framework for User-Led Content Creation*. Presented at the Creativity & Cognition Conference, Washington.
Bruns, A. (2008a) 'The Future Is User-Led: The Path towards Widespread Produsage'. *Fibreculture* [online] (11), available from http://eleven.fibreculturejournal.org/fcj-066-the-future-is-user-led-the-path-towards-widespread-produsage/.
Bruns, A. (2008b) *Blogs, Wikipedia, Second Life, and Beyond: From Production to Produsage*. New York: Peter Lang.
Bucher, T. (2012) 'Want to Be on the Top? Algorithmic Power and the Threat of Invisibility on Facebook'. *New Media & Society* 14 (7): 1164–80.
Cadwalladr, C. and Graham-Harrison, E. (2018) 'Revealed: 50 Million Facebook Profiles Harvested for Cambridge Analytica in Major Data Breach'. *The Guardian* [online] 17 March, available from https://www.theguardian.com/news/2018/mar/17/cambridge-analytica-facebook-influence-us-election [1 March 2020].
Campaign Choirs Writing Collective (2018) *Singing for Our Lives: Stories from the Street Choirs*. Bristol: HammerOn Press.
Caraway, B. (2011) 'Audience Labor in the New Media Environment: A Marxian Revisiting of the Audience Commodity'. *Media, Culture & Society* 33 (5): 693–708.
Caraway, B. (2016) 'Crisis of Command: Theorizing Value in New Media'. *Communication Theory* 26 (1) 64–81.
Casemajor, N., Couture, S., Delfin, M., Goerzen, M. and Delfanti, A. (2015) 'Non-Participation in Digital Media: Toward a Framework of Mediated Political Action'. *Media, Culture & Society* 37 (6): 850–66.
Cellan-Jones, R. (2012) *Who 'Likes' My Virtual Bagels?* [online] available from http://www.bbc.co.uk/news/technology-18819338 [27 February 2017].
Chandler, D. and Munday, R. (2016a) *'Spam'* [online] available from http://www.oxfordreference.com/view/10.1093/acref/9780191803093.001.0001/acref-9780191803093 [1 June 2017].

Chandler, D. and Munday, R. (2016b) *'Trolling'* [online] available from http://www.oxfordreference.com/view/10.1093/acref/9780191803093.001.0001/acref-9780191803093 [7 June 2017].

Chrysagis, E. (2016) 'The Visible Evidence of DiY Ethics: Music, Publicity and Technologies of (In)Visibility in Glasgow'. *Visual Culture in Britain* 17 (3): 290–310.

Clarke, J. (2006) 'The Skinheads and the Magical Recovery of Community'. In *Resistance through Rituals: Youth Subcultures in Post-War Britain*, 2nd edn, edited by Hall, S. and Jefferson, T. London: Routledge, 80–8.

Cohen, S. (1991) *Rock Culture in Liverpool: Popular Music in the Making*. Oxford: Clarendon Press.

Cohen, J. E. (2012) *Configuring the Networked Self: Law, Code, and the Play of Everyday Practice*. New Haven: Yale University Press.

Cotter, K. (2019) 'Playing the Visibility Game: How Digital Influencers and Algorithms Negotiate Influence on Instagram'. *New Media & Society* 21 (4): 895–913.

Crenshaw, K. (1991) 'Mapping the Margins: Intersectionality, Identity Politics, and Violence against Women of Color'. *Stanford Law Review* 43 (6): 1241–99.

Croteau, D. (2006) 'The Growth of Self-Produced Media Content and the Challenge to Media Studies'. *Critical Studies in Media Communication* 23 (4): 340–4.

Csikszentmihalyi, M. (1990) *Flow: The Psychology of Optimal Experience*. London: HarperCollins.

Culton, K. R. and Holtzman, B. (2010) 'The Growth and Disruption of a "Free Space": Examining a Suburban Do It Yourself (DIY) Punk Scene'. *Space and Culture* 13 (3): 270–84.

Dale, P. (2012) *Anyone Can Do It: Empowerment, Tradition and the Punk Underground*. London: Routledge.

Darms, Lisa and Fateman, J. (2013) *The Riot Grrrl Collection*, edited by Darms, L. New York: Feminist Press.

Davyd, M. and Whitrick, B. (2015) *Understanding Small Music Venues*. London: Music Venue Trust.

Dawes, L. (2013) *Why I Was Never a Riot Grrrl* [online] available from https://bitchmedia.org/post/why-i-was-never-a-riot-grrl [18 September 2016].

Dean, J. (2010) *Blog Theory: Feedback and Capture in the Circuits of Drive*. Cambridge: Polity.

Deleuze, G. (1992) 'Postscript on the Societies of Control'. *October* 59: 3–7.

Devine, K. (2019) *Decomposed: The Political Ecology of Music*. Cambridge, MA: MIT Press.

DIY Access Guide (2017) Attitude Is Everything, available from http://www.attitudeiseverything.org.uk/diyaccessguide [5 February 2020].

DIY Space for London (2016) *You Can't Be What You Can't See: DIY Performers Talk Identity, Gender and Music* [online] available from https://diyspaceforlondon.org/event/you-cant-be-what-you-cant-see/ [19 July 2017].

Dolan, E. I. (2010) '"… This Little Ukulele Tells the Truth": Indie Pop and Kitsch Authenticity'. *Popular Music* 29 (3): 457–69.

Downes, J. (2009) *DIY Queer Feminist (Sub)Cultural Resistance in the UK*. Doctoral dissertation. University of Leeds.

Downes, J. (2012) 'The Expansion of Punk Rock: Riot Grrrl Challenges to Gender Power Relations in British Indie Music Subcultures'. *Women's Studies* 41 (2): 204–37.

Downes, J., Breeze, M. and Griffin, N. (2013) 'Researching DIY Cultures: Towards a Situated Ethical Practice for Activist-Academia'. *Graduate Journal of Social Science* 10 (3): 100–24.

Downey, J. and Toynbee, J. (2016) 'Ideology: Towards Renewal of a Critical Concept'. *Media, Culture & Society* 38 (8): 1261–71.

Dryhurst, M. (2015) *Saga v1.0* [online] available from https://accessions.org/article/saga-v1-0/ [8 September 2017].

Duffy, B. E. (2017) *(Not) Getting Paid to Do What You Love: Gender, Social Media, and Aspirational Work*. New Haven, CT: Yale University Press.

Duffy, B. E., Poell, T. and Nieborg, D. B. (2019) 'Platform Practices in the Cultural Industries: Creativity, Labor, and Citizenship'. *Social Media + Society* 5 (4): 1–8.

Duncombe, S. (2002) *Cultural Resistance Reader*. London: Verso.

Duncombe, S. (2008) *Notes from the Underground: Zines and the Politics of Underground Culture*, 2nd edn. Bloomington, IN: Microcosm.

Duncombe, S. (2011) *White Riot: Punk Rock and the Politics of Race*. London: Verso.

Duncombe, S. (2017) 'Resistance'. In *Keywords for Media Studies*, edited by Ouellette, L. and Gray, J. New York: New York University Press, 176–9.

Dunn, K. (2016) *Global Punk: Resistance and Rebellion in Everyday Life*. London: Bloomsbury.

Dunn, K. and Farnsworth, M. S. (2012) '"We ARE the Revolution": Riot Grrrl Press, Girl Empowerment, and DIY Self-Publishing'. *Women's Studies* 41 (2): 136–57.

Dyer-Witheford, N. (1999) *Cyber-Marx: Cycles and Circuits of Struggle in High-Technology Capitalism*. Urbana, IL: University of Illinois Press.

Enzensberger, H. M. (1974) *The Consciousness Industry: On Literature, Politics and the Media*. New York: Seabury Press.

Ewens, H. (2019) *Fangirls: Scenes from Modern Music Culture*. London: Quadrille.

Faris, M. (2004) '"That Chicago Sound": Playing with (Local) Identity in Underground Rock'. *Popular Music and Society* 27 (4): 429–54.

Fatsis, L. (2019) 'Policing the Beats: The Criminalisation of UK Drill and Grime Music by the London Metropolitan Police': *The Sociological Review* 67 (6): 1300–16.

Fife, K. (2019) 'Not For You? Ethical Implications of Archiving Zines', *Punk & Post-Punk*, 8(2): 227–42.

Finn, E. (2017) *What Algorithms Want: Imagination in the Age of Computing*. Cambridge, MA: MIT Press.

Finnegan, R. (1988) *The Hidden Musicians: Music-Making in an English Town*. Cambridge: Cambridge University Press.

First Timers (2017) *First Timers* [online] available from http://www.firsttimers.org/ [19 July 2017].

Fisher, M. (2009) *Capitalist Realism: Is There No Alternative?* Ropley: Zero Books.

Fraser, N. (1990) 'Rethinking the Public Sphere: A Contribution to the Critique of Actually Existing Democracy'. *Social Text* (25/26): 56–80.

Fraser, N. (2000) 'Rethinking Recognition'. *New Left Review*, May–June (3): 107–20.

Fraser, N. (2003) 'Social Justice in the Age of Identity Politics: Redistribution, Recognition, and Participation'. In *Redistribution or Recognition?: A Political-Philosophical Exchange*, edited by Fraser, N. and Honneth, A. London: Verso, 7–109.

Fraser, N. (2009) 'Feminism, Capitalism, and the Cunning of History'. *New Left Review* March–Apr (56): 97–117.

Frith, S. (1996) *Performing Rites: On the Value of Popular Music*. Cambridge, MA: Harvard.

Frith, S. (2004) 'Afterword'. In *After Subculture: Critical Studies in Contemporary Youth Culture*, edited by Bennett, A. and Kahn-Harris, K. Basingstoke: Palgrave Macmillan, 173–7.

Frith, S. (2007) 'Towards an Aesthetic of Popular Music'. In *Taking Popular Music Seriously*. Aldershot: Ashgate, 257–73.

Fuchs, C. (2012) 'Dallas Smythe Today – The Audience Commodity, the Digital Labour Debate, Marxist Political Economy and Critical Theory. Prolegomena to a Digital Labour Theory of Value'. *TripleC* 10 (2): 692–740.

Fuchs, C. and Dyer-Witheford, N. (2013) 'Karl Marx @ Internet Studies'. *New Media & Society* 15 (5): 782–96.

Galloway, A. R. and Thacker, E. (2007) *The Exploit: A Theory of Networks*. Minneapolis, MN: University of Minnesota Press.

Gang of Four (1979) *Entertainment!* [LP]. EMI, EMC3313.

Garnham, N. (1990) *Capitalism and Communication: Global Culture and the Economics of Information*. London: Sage.

Garrison, E. K. (2000) 'U.S. Feminism-Grrrl Style! Youth (Sub) Cultures and the Technologics of the Third Wave'. *Feminist Studies* 26 (1): 141–70.

Geffen, S. (2020) 'No Shape: How Tech Helped Musicians Melt the Gender Binary. *The Guardian* [online] 7 April, available from https://www.theguardian.com/news/2018/mar/17/cambridge-analytica-facebook-influence-us-election [4 May 2020].

Gerlitz, C. and Helmond, A. (2013) 'The Like Economy: Social Buttons and the Data-Intensive Web'. *New Media & Society* 15 (8): 1348–65.

Gerrard, Y. (2017) '"It's a Secret Thing": Digital Disembedding through Online Teen Drama Fandom'. *First Monday* [online] 22 (8). Available from http://www.ojphi.org/ojs/index.php/fm/article/view/7877/6514.

Gibson-Graham, J. K. (2006) *A Postcapitalist Politics*. Minneapolis: University of Minnesota Press.

Giddens, A. (1991) *Modernity and Self-Identity: Self and Society in the Late Modern Age*. Stanford, CA: Stanford University Press.

Gill, A. (1978) 'Wire: Limit Club, Sheffield'. *New Musical Express*, 13 May.

Gillespie, T. (2010) 'The Politics of "Platforms"'. *New Media & Society* 12 (3): 347–64.

Gillespie, T. (2017) *The Platform Metaphor, Revisited* [online] available from https://www.hiig.de/en/blog/the-platform-metaphor-revisited/ [1 September 2017].

Gillespie, T. (2018) *Custodians of the Internet: Platforms, Content Moderation, and the Hidden Decisions That Shape Social Media*. New Haven, CT: Yale University Press.

Goldhaber, M. H. (1997) *The Attention Economy and the Net* [online] available from https://journals.uic.edu/ojs/index.php/fm/article/download/519/440?inline=1 [1 March 2020].

Golpushnezhad, E. (2018) 'Untold Stories of DIY/Underground Iranian Rap Culture: The Legitimization of Iranian Hip-Hop and the Loss of Radical Potential'. *Cultural Sociology* 12 (2): 260–75.

Gosden, E. (2016) *Student Accused of Violating University 'Safe Space' by Raising Her Hand* [online] available from http://www.telegraph.co.uk/news/2016/04/03/student-accused-of-violating-university-safe-space-by-raising-he/ [27 February 2017].

Gottlieb, J. and Wald, G. (1994) 'Smells Like Teen Spirit: Riot Grrrls, Revolution and Women in Independent Rock'. In *Microphone Fiends: Youth Music and Youth Culture*, edited by Ross, A. and Rose, T. New York: Routledge, 250–74.

Gracyk, T. (2012) 'Kids're Forming Bands: Making Meaning in Post-Punk'. *Punk & Post-Punk* 1 (1): 73–85.

Graham, S. (2016) *Sounds of the Underground: A Cultural, Political, and Aesthetic Mapping of Underground and Fringe Music*. Tracking pop. Ann Arbor: University of Michigan Press.

Gray, K. L. (2012) 'Intersecting Oppressions and Online Communities'. *Information, Communication & Society* 15 (3): 411–28.

Gross, S. A. and Musgrave, G. (2016) *Can Music Make You Sick? Music and Depression: A Study into the Incidence of Musicians' Mental Health*. London: MusicTank.

Grosser, B. (2014) 'What Do Metrics Want? How Quantification Prescribes Social Interaction on Facebook'. *Computational Culture* 4: 1–38.

Guerra, P. (2018) 'Raw Power: Punk, DIY and Underground Cultures as Spaces of Resistance in Contemporary Portugal'. *Cultural Sociology* 12 (2): 241–59.

Gurak, L. J. (2001) *Cyberliteracy: Navigating the Internet with Awareness*. New Haven: Yale University Press.

Gurman, D. (2009) 'Why Lakoff Still Matters: Framing the Debate on Copyright Law and Digital Publishing'. *First Monday* 14 (6).

Hancox, D. (2019) *Inner City Pressure: The Story of Grime*. London: W. Collins.

Hardt, M. and Negri, A. (2005) *Multitude: War and Democracy in the Age of Empire*. Penguin: London.

Harrison, A. K. (2006) '"Cheaper than a CD, Plus We Really Mean It": Bay Area Underground Hip Hop Tapes as Subcultural Artefacts'. *Popular Music* 25 (2): 283–301.

Harvie, D. (2004) 'Commons and Communities in the University: Some Notes and Some Examples'. *The Commoner* (8), Autumn/Winter.

Hauben, M. (1997) 'The Computer as a Democratizer'. In *Netizens: On The History and Impact of Usenet and The Internet* [online], edited by Hauben, R. and Hauben, M. Los Alamitos: IEEE Computer Society Press, 315–20. Available from http://www.columbia.edu/~rh120/ch106.x18 [18 February 2020].

Hayler, R. (2015) *What I Mean by the Term 'No-Audience Underground'*. 2015 Remix [online] available from https://radiofreemidwich.wordpress.com/2015/06/14/what-i-mean-by-the-term-no-audience-underground-2015-remix/ [12 July 2017].

Haythornthwaite, C. (2002) 'Strong, Weak, and Latent Ties and the Impact of New Media'. *Information Society* 18 (5): 385–401.

Haythornthwaite, C. (2005) 'Social Networks and Internet Connectivity Effects'. *Information, Communication & Society* 8 (2): 125–47.

Hearn, A. (2017) 'Producing "Reality": Branded Content, Branded Selves, Precarious Futures'. In *A Companion to Reality Television*, edited by Ouellette, L. Wiley Blackwell, 437–55.

Hearn, A. and Banet-Weiser, S. (2020) 'The Beguiling: Glamour in/as Platformed Cultural Production'. *Social Media + Society* 6 (1): 2056305119898779.

Hendriks, C. M., Duus, S. and Ercan, S. A. (2016) 'Performing Politics on Social Media: The Dramaturgy of an Environmental Controversy on Facebook'. *Environmental Politics* 25 (6): 1102–25.

Herstand, A. (2016) *How To Make It in the New Music Business*. New York: Liveright.

Hesmondhalgh, D. (1997) 'Post-Punk's Attempt to Democratise the Music Industry: The Success and Failures of Rough Trade'. *Popular Music* 16 (3): 255–74.

Hesmondhalgh, D. (2005) 'Subcultures, Scenes or Tribes? None of the Above'. *Journal of Youth Studies* 8 (1): 21–40.

Hesmondhalgh, D. (2010) 'User-Generated Content, Free Labour and the Cultural Industries Theory and Politics in Organization'. *Ephemera: Theory & Politics in Organization* 10 (3/4): 267–84.

Hesmondhalgh, D. (2013) *Why Music Matters*. West Sussex: Wiley-Blackwell.

Hesmondhalgh, D. (2019) *The Cultural Industries*, 4th edn. London: Sage.

Hesmondhalgh, D. and Baker, S. (2011) *Creative Labour: Media Work in Three Cultural Industries*. London: Routledge.

Hesmondhalgh, D. and Meier, L. (2015) 'Popular Music, Independence and the Concept of the Alternative in Contemporary Capitalism'. In *Media Independence: Working with Freedom or Working for Free?* edited by Bennett, J. and Strange, N. New York: Routledge, 94–116.

Hesmondhalgh, D., Jones, E. and Rauh, A. (2019) 'SoundCloud and Bandcamp as Alternative Music Platforms'. *Social Media + Society* 5: 1–13.

Hines, S. (2019) 'The Feminist Frontier: On Trans and Feminism'. *Journal of Gender Studies* 28 (2): 145–57.

Horkheimer, M. and Adorno, T. W. (2002) 'The Culture Industry: Enlightenment as Mass Deception'. *Dialectic of Enlightenment*. London: Verso, 41–72.

Hosie, R. (2017) *Millennials Are Struggling at Work Because Their Parents 'Gave Them Medals for Coming Last'* [online] available from http://www.independent.co.uk/life-style/millennials-struggling-work-careers-because-their-parents-gave-them-medals-for-coming-last-simon-a7537121.html [27 February 2017].

Howard-Spink, S. (2004) 'Grey Tuesday, Online Cultural Activism and the Mash-Up of Music and Politics'. *First Monday* 9 (10).

Hui, Y. and Halpin, H. (2013) 'Collective Individuation: The Future of the Social'. In *Unlike Us Reader*, edited by Lovink, G. and Rasch, M. Amsterdam: Institute of Network Cultures, 103–16.

Jakobsson, P. (2010) 'Cooperation and Competition in Open Production'. *Platform: Journal of Media and Communication* 2 (1): 106–19.

Jenkins, H. (2008) *The Participation Gap* [online] available from http://www.nea.org/home/15468.htm.

Jenkins, H., Clinton, K., Purushotma, R., Robison, A. J. and Weigel, M. (2006) *Confronting the Challenges of Participatory Culture: Media Education for the 21st Century* [online] Chicago, IL. Available from https://www.macfound.org/media/article_pdfs/JENKINS_WHITE_PAPER.PDF.

Jhally, S. (1982) 'Probing the Blindspot: The Audience Commodity'. *Canadian Journal of Political and Social Theory* 6 (1–2): 204–10.

John, N. (2013) 'The Social Logics of Sharing'. *The Communication Review* 16 (3): 113–31.

John, N. (2017) 'Sharing – from Ploughshares to File Sharing and Beyond' [online]. *HIIG Digital Society Blog*, https://www.hiig.de/en/sharing-from-ploughshares-to-file-sharing-and-beyond/ [1 February 2020].

Jones, E. (2019) 'What Does Facebook "Afford" Do-It-Yourself Musicians? Considering Social Media Affordances as Sites of Contestation'. *Media, Culture & Society* 42 (2): 277–92.

Jones, E. (2021) 'DIY and Popular Music: Mapping an Ambivalent Relationship across Three Historical Case Studies', Popular Music and Society, 44(1).

Jouhki, J., Lauk, E., Penttinen, M., Sormanen, N., and Uskali, T. (2016) 'Facebook's Emotional Contagion Experiment as a Challenge to Research Ethics'. *Media and Communication* 4 (4): 75.

Jurgenson, N. (2010) 'Review of Ondi Timoner's "We Live in Public"'. *Surveillance & Society* 8 (3): 377.

Kahn-Harris, K. (2004) 'Unspectacular Subculture? Transgression and Mundanity in the Global Extreme Metal Scene'. In *After Subculture: Critical Studies in Contemporary Youth Culture*, edited by Bennett, A. and Kahn-Harris, K. Basingstoke: Palgrave Macmillan, 107–18.

Kanai, A. (2015a) 'DIY Culture'. In *The Routledge Companion to Remix Studies*, edited by Navas, E., Gallagher, O. and Burrough, X. London: Routledge, 125–34.

Kanai, A. (2015b) 'WhatShouldWeCallMe? Self-Branding, Individuality and Belonging in Youthful Femininities on Tumblr'. *M/C Journal* [online] 18 (1). Available from http://journal.media-culture.org.au/index.php/mcjournal/rt/printerFriendly/936/0 [1 March 2020].

Kanai, A. (2019) *Gender and Relatability in Digital Culture: Managing Affect, Intimacy, and Value*. Cham, Switzerland: Palgrave Macmillan.

Kang, J. (2000) 'Cyber-Race'. *Harvard Law Review* 113 (5): 1130–208.

Kaun, A. and Stiernstedt, F. (2014) 'Facebook Time: Technological and Institutional Affordances for Media Memories'. *New Media & Society* 16 (7): 1154–68.

Keightley, K. (2001) 'Reconsidering Rock'. In *The Cambridge Companion to Pop and Rock*, edited by Frith, S. Cambridge: Cambridge University Press, 109–42.

Kennedy, H. (2016) *Post, Mine, Repeat: Social Media Data Mining Becomes Ordinary*. London: Palgrave Macmillan.

Kenney, Moira. (2001) *Mapping Gay L.A.: The Intersection of Place and Politics*. Philadelphia: Temple University Press.

Kim, J. (2012) 'The Institutionalization of YouTube: From User-Generated Content to Professionally Generated Content'. *Media, Culture & Society* 34 (1): 53–67.

Kinder, K. (2016) *DIY Detroit: Making Do in a City without Services*. Minneapolis: University of Minnesota Press.

King, J. (2019) *Activism, Identity Politics, and Pop's Great Awokening* [online] available from https://pitchfork.com/features/article/2010s-pops-great-awokening-black-lives-matter-beyonce-kendrick-lamar-solange/.

Klein, B. (2020) *Selling Out: Culture, Commerce and Popular Music*. New York: Bloomsbury.

Klein, B., Meier, L.M. and Powers, D. (2017) 'Selling Out: Musicians, Autonomy, and Compromise in the Digital Age'. *Popular Music and Society* 40 (2): 222–38.

Koch, K. (2006) *Don't Need You: The Herstory of Riot Grrrl* [DVD]. Urban Cowgirl Productions.

Kruse, H. (1993) 'Subcultural Identity in Alternative Music Culture'. *Popular Music* 12 (1): 33–41.

Kuehn, K. and Corrigan, T. F. (2013) 'Hope Labor: The Role of Employment Prospects in Online Social Production'. *The Political Economy of Communication* 1 (1): 9–25.

Laing, D. (1985) *One Chord Wonders: Power and Meaning in Punk Rock*. Maidenhead: Open University Press.

Langlois, G. (2013) 'Social Media, or Towards a Political Economy of Psychic Life'. In *Unlike Us Reader*, edited by Lovink, G. and Rasch, M. Amsterdam: Institute of Network Cultures, 50–61.

Lee, M. (2011) 'Google Ads and the Blindspot Debate'. *Media Culture & Society* 33 (3): 433–47.

Levmore, S. (2010) 'The Internet's Anonymity Problem'. In *The Offensive Internet: Speech, Privacy and Reputation*, edited by Nussbaum, M. and Levmore, S. Cambridge, MA: Harvard University Press, 51–70.

Light, B., Burgess, J., & Duguay, S. (2018). 'The Walkthrough Method: An Approach to the Study of Apps'. *New Media & Society* 20 (3): 881–900.

Litt, E. and Hargittai, E. (2016) 'The Imagined Audience on Social Network Sites'. *Social Media + Society* January–March, 1–12.

Livant, B. (1978) 'The Audience Commodity: On the Blindspot Debate'. *Canadian Journal of Political and Social Theory* [online] 1 (1). Available from https://mmduvic.ca/index.php/ctheory/article/view/13793.

Live Music Awards (2015) *Winners 2015* [online] available from http://www.livemusicawards.co.uk/winners-2015/ [29 October 2017].

Loten, A., Janofsky, A. and Albergotti, R. (2014) *New Facebook Rules Will Sting Entrepreneurs* [online] available from https://www.wsj.com/articles/new-facebook-rules-will-sting-entrepreneurs-1417133694 [8 June 2017].

Lueg, C. and Fisher, D. (2012) *From Usenet to CoWebs: Interacting with Social Information Spaces*. London: Springer Science & Business Media.

Luxemburg, R. (2006) *Revolt or Revolution and Other Writings*. Mineola, NY: Dover.

Mack, Z. (2019) *How Streaming Affects the Lengths of Songs* [online]. *The Verge*. Available from https://www.theverge.com/2019/5/28/18642978/music-streaming-spotify-song-length-distribution-production-switched-on-pop-vergecast-interview [1 March 2020].

Manzerolle, V. (2010) 'Mobilizing the Audience Commodity: Digital Labour in a Wireless World'. *Ephemera* [online] 10 (3). Available from http://scholar.uwindsor.ca/communicationspub.

Manzerolle, V. and McGuigan, L. (2014) *The Audience Commodity in a Digital Age: Revisiting a Critical Theory of Commercial Media*. New York: Peter Lang.

Marcuse, H. (1991) *One-Dimensional Man: Studies in the Ideology of Advanced Industrial Society*. London: Routledge.

Marder, B., Joinson, A., Shankar, A. and Houghton, D. (2016a) 'The Extended "Chilling" Effect of Facebook: The Cold Reality of Ubiquitous Social Networking'. *Computers in Human Behavior* 60: 582–92.

Marder, B., Slade, E., Houghton, D. and Archer-Brown, C. (2016b) '"I Like Them, but Won't 'Like' Them": An Examination of Impression Management Associated with Visible Political Party Affiliation on Facebook'. *Computers in Human Behavior* 61: 280–7.

Marwick, A. E. (2015) 'Instafame: Luxury Selfies in the Attention Economy'. *Public Culture* 27 (1): 137–60.

Marwick, A. E. and boyd, d. (2010) 'I Tweet Honestly, I Tweet Passionately: Twitter Users, Context Collapse, and the Imagined Audience'. *New Media & Society* 13 (1): 114–33.

Marx, K. (1976) *Capital, Vol. 1*. London: Penguin.

Marx, K. (2000) 'Economic and Philosophical Manuscripts'. In *Early Writings*. London: Penguin, 279–400.

Matarasso, F. (1997) *Use or Ornament? The Social Impact of Participation in the Arts*. Stroud: Comedia.

Mattern, M. (1994) *Acting in Concert. Music, Community And Political Action*. London: Rutgers University Press.

Mayer-Schönberger, V. and Cukier, K. (2013) *Big Data: A Revolution That Will Transform How We Live, Work, and Think*. New York: Houghton Mifflin Harcourt.

Mayo, C. (2010) 'Incongruity and Provisional Safety: Thinking Through Humor'. *Studies in Philosophy and Education* 29 (6): 509–21.

Mazierska, E., Gillon, L. and Rigg, T. (2018) 'Introduction: The Future of and through Music'. In *Popular Music in the Post-Digital Age*, edited by Mazierska, E., Gillon, L. and Rigg, T. London: Bloomsbury, 1–31.

McChesney, R. (2013) *Digital Disconnect: How Capitalism Is Turning the Internet against Democracy*. London: The New Press.

McKay, G. (1998) *DiY Culture: Party and Protest in Nineties Britain*. London: Verso.

McMillan Cottom, T. (2015) '"Who Do You Think You Are?": When Marginality Meets Academic Microcelebrity'. *Ada: A Journal of Gender, New Media, and Technology* [online] 7. Available from https://adanewmedia.org/2015/04/issue7-mcmillancottom/ [8 June 2019].

McQuaid, I. (2015) *XTC, Functions On The Low – A Lost Grime Classic* [online] available from https://www.redbull.com/gb-en/stormzy-shut-up-producer-xtc-interview [18 February 2020].

McRobbie, A. (2002) 'Clubs To Companies: Notes on the Decline of Political Culture in Speeded Up Creative Worlds'. *Cultural Studies* 16 (4): 516–31.

Meads, N. (2016) *Superb @scrittipolitti 1980 DIY 'Release a Record' Guide, Stumbled upon Researching #SmallWonderRecords Exhibition* [online] available from https://twitter.com/musiclikedirt/status/756564212760084484?lang=en-gb [5 May 2017].

Meier, L. M. (2017) *Popular Music as Promotion: Music and Branding in the Digital Age*. Cambridge: Polity.

Mendick, H. (2013) *Online Communication in the Age of Fast Academia* [online] available from http://www.celebyouth.org/the_internet_and_fast_academia/ [2 September 2017].
Miège, B. (1989) *The Capitalization of Cultural Production*. New York: International General.
Miller, D. (2001) *Consumption: Critical Concepts in the Social Sciences, Vol. 1.* London: Routledge.
Moore, R. (2004) 'Postmodernism and Punk Subculture: Cultures of Authenticity and Deconstruction'. *The Communication Review* 7 (3): 305–27.
Morris, J. W. (2015) *Selling Digital Music, Formatting Culture*. Oakland: University of California Press.
Mullaney, J. L. (2007) '"Unity Admirable but Not Necessarily Heeded": Going Rates and Gender Boundaries in the Straight Edge Hardcore Music Scene'. *Gender & Society* 21 (3): 384–408.
Murdock, G. (1978) 'Blindspots about Western Marxism: A Reply to Dallas Smythe'. *CTheory* 2 (2): 109–15.
Myslik, W. D. (1996) 'Renegotiating the Social Identities of Place'. In *BodySpace: Destabilising Geographies of Gender and Sexuality*, edited by Duncan, N. London: Routledge, 157–69.
Nafus, D. (2016) 'The Domestication of Data: Why Embracing Digital Data Means Embracing Bigger Questions'. *Ethnographic Praxis in Industry Conference Proceedings* 2016 (1): 384–99.
Needham, A. (2018) 'Neil Tennant: "Sometimes I Think, Where's the Art, the Poetry in All This?"' *Guardian* [online] available from https://www.theguardian.com/music/2018/oct/21/neil-tennant-pet-shop-boys-collection-lyrics.
Nguyen, M. T. (2012) 'Riot Grrrl, Race, and Revival'. *Women & Performance: A Journal of Feminist Theory* 22 (2–3): 173–96.
Nieborg, D. B. and Helmond, A. (2018) 'The Political Economy of Facebook's Platformization in the Mobile Ecosystem: Facebook Messenger as a Platform Instance'. *Media, Culture & Society* 41 (2): 196–218.
Nieborg, D. B. and Poell, T. (2018) 'The Platformization of Cultural Production: Theorizing the Contingent Cultural Commodity'. *New Media & Society* 20 (11): 4275–92.
Noble, S. U. (2018) *Algorithms of Oppression: How Search Engines Reinforce Racism*. New York: New York University Press.
O'Brien, L. (2012) 'Can I Have a Taste of Your Ice Cream ?' *Punk & Post-Punk* 1 (1): 27–40.
Ogg, A. (2009) *Independence Days: The Story of UK Independent Record Labels*. London: Cherry Red Books.
Packard, V. (1957) *The Hidden Persuaders*. London: Longmans.
Papacharissi, Z. (2002) 'The Virtual Sphere: The Internet as a Public Sphere'. *New Media & Society* 4 (1): 9–27.
Pariser, Eli. (2011) *The Filter Bubble: What the Internet Is Hiding from You*. London: Penguin.
Patelis, K. (2013) 'Political Economy and Monopoly Abstractions: What Social Media Demand'. In *Unlike Us Reader*, edited by Lovink, G. and Rasch, M. Amsterdam: Institute of Network Cultures, 117–26.

Pearce, R. and Lohman, K (2019) 'De/constructing DIY Identities in a Trans Music Scene'. *Sexualities* 22 (1–2): 97–113.
Pelly, L. (2019) 'The Antisocial Network'. *Logic* [online] 1 January. Available from https://logicmag.io/play/the-antisocial-network/ [10 March 2020].
Petre, C., Duffy, B. E. and Hund, E. (2019) '"Gaming the System": Platform Paternalism and the Politics of Algorithmic Visibility'. *Social Media + Society* 5 (4): 1–12.
Phillips, S. (2017) 'The Bands Taking British Punk Back to Its Multicultural Roots'. Available from https://www.vice.com/en_uk/article/padjev/decolonise-fest-uk-punk-nekra-sacred-paws-fight-rosa.
Pitcan, P., Marwick A. E., and boyd, d. (2018) 'Performing a Vanilla Self: Respectability Politics, Social Class, and the Digital World'. *Journal of Computer-Mediated Communication* 23 (3): 163–79.
Piepmeier, A. (2009) *Girl Zines: Making Media, Doing Feminism*. London: New York University Press.
Prior, N. (2015) 'Beyond Napster: Popular Music and the Normal Internet'. In *The Handbook of Popular Music*, edited by Bennett, A. and Waksman, S. London: Sage, 493–507.
Purdue, D., Dürrschmidt, J., Jowers, P. and O'Doherty, R. (1997) 'DIY Culture and Extended Miliuex: LETS, Veggie Boxes and Festivals'. *Sociological Review* 45 (4): 645–67.
Putnam, R. D. (2000) *Bowling Alone: The Collapse and Revival of American Community*. New York: Simon & Schuster.
Quader, S. B. and Redden, G. (2015) 'Approaching the Underground: The Production of Alternatives in the Bangladeshi Metal Scene'. *Cultural Studies* 29 (3): 401–24.
Radway, J. (2016) 'Girl Zine Networks, Underground Itineraries, and Riot Grrrl History: Making Sense of the Struggle for New Social Forms in the 1990s and beyond'. *Journal of American Studies* 50 (1): 1–31.
Reia, J. (2014) 'Napster and beyond: How Online Music Can Transform the Dynamics of Musical Production and Consumption in DIY Subcultures'. *First Monday* [online] 19 (10). Available from http://journals.uic.edu/ojs/index.php/fm/article/view/5552 [18 February 2020].
Rentschler, C. A. (2017) 'Bystander Intervention, Feminist Hashtag Activism, and the Anti-Carceral Politics of Care'. *Feminist Media Studies* 17 (4): 565–84.
Resonate (2015) *Stream-to-Own* [online] available from https://resonate.is/stream2own/ [8 September 2017].
Resonate (n.d.) *Own the Future* [online] available from https://resonate.is/inviting-you-to-own-the-future/ [2 September 2017].
Rey, P. J. (2012) 'Alienation, Exploitation, and Social Media'. *American Behavioral Scientist* 56 (4): 399–420.
Reynolds, S. (2005) *Rip It Up and Start Again: Post-Punk 1978–84*. London: Faber and Faber.
Reynolds, S. (2012) '*DIY Conference: An Update with Simon Reynolds*'. Incubate Festival [Film] available from https://vimeo.com/49913417 [22 July 2017].
Ritzer, G. and Jurgenson, N. (2010) 'Production, Consumption, Prosumption: The Nature of Capitalism in the Age of the Digital "Prosumer"'. *Journal of Consumer Culture* 10 (1): 13–36.

Roberts, S. T. (2019) *Behind the Screen: Content Moderation in the Shadows of Social Media*. New Haven, London: Yale University Press.

Robertson, A. (2019) *Should We Treat Incels as Terrorists?* [online] available from https://www.theverge.com/2019/10/5/20899388/incel-movement-blueprint-toronto-attack-confession-gender-terrorism [1 March 2020].

Rosen, P. (1997) 'It Was Easy, It Was Cheap, Go and Do It!' Technology and Anarchy in the UK Music Industry'. In *Twenty-First Century Anarchism: Unorthodox ideas for a New Millennium*, edited by Purkis, J. & Bowen, J. New York: Cassell Press, 99–116.

Rosen, J. (2006) *The Perils of Poptimism: Does Hating Rock Make You a Music Critic?* [online] available from http://www.slate.com/articles/arts/music_box/2006/05/the_perils_of_poptimism.html [27 February 2017].

Ruckenstein, M. and Pantzar, M. (2017) 'Beyond the Quantified Self: Thematic Exploration of a Dataistic Paradigm'. *New Media & Society* 19 (3): 401–18.

Ryan, C. (2015) *'Against the Rest': Fanzines and Alternative Music Cultures in Ireland*. Ireland: University of Limerick.

Sahim, S. (2015) *The Unbearable Whiteness of Indie* [online] available from http://pitchfork.com/thepitch/710-the-unbearable-whiteness-of-indie/.

Scarborough, R. C. (2017) 'Making It in a Cover Music Scene: Negotiating Artistic Identities in a "Kmart-Level Market"'. *Sociological Inquiry* 87 (1): 153–78.

Schilt, K. (2003a) '"A Little Too Ironic": The Appropriation and Packaging of Riot Grrrl Politics by Mainstream Female Musicians'. *Popular Music and Society* 26 (1): 5–16.

Schilt, K. (2003b) '"I'll Resist with Every Inch and Every Breath": Girls and Zine Making as a Form of Resistance'. *Youth Society* 35 (1): 71–97.

Schoop, M. E. (2017) *Independent Music and Digital Technology in the Philippines*. London: Routledge.

Scritti Politti (1985) *Cupid & Psyche 85* [LP]. Virgin, V2350.

Selzer, D. (2012) *Xerox Music Is Here to Stay* [online] available from http://swingsetmagazine.com/2012/06/xerox_music_is_here_to_stay/ [21 December 2016].

Sennett, R. (2009) *The Craftsman*. London: Penguin.

Shank, B. (1994) *Dissonant Identities: The Rock'n'roll Scene in Austin, Texas*. Hanover, NH: University Press of New England.

Silbaugh, K. (2011) 'Testing as Commodification'. *Washington University Journal of Law & Policy* 35: 309–36.

Simon, P. (1983) 'When Numbers Get Serious'. Hearts and Bones [LP]. Warner Bros, 9 23942-1.

Sinker, D. (ed.) (2001) *We Owe You Nothing: Punk Planet, the Collected Interviews*. New York: Akashic Books.

Smith, M. S. and Giraud-Carrier, C. (2010) 'Bonding vs. Bridging Social Capital: A Case Study in Twitter'. In *2010 IEEE Second International Conference on Social Computing*, held August 2010. Washington DC: IEEE Computer Society, 385–92.

Smythe, D. (1977) 'Communications: Blindspot of Western Marxism'. *CTheory* [online] available from http://mmduvic.ca/index.php/ctheory/article/view/13715 [12 August 2017].

Smythe, D. (1978) 'Rejoinder to Graham Murdock'. *CTheory* [online] available from http://mmduvic.ca/index.php/ctheory/article/view/13745.

Smythe, D. (1981) *Dependency Road: Communications, Capitalism, Consciousness, and Canada* [online]. Norwood, NJ: Ablex.

Solnit, R. (2006) *Hope in the Dark: Untold Histories, Wild Possibilities*. New York: Nation Books.

Srnicek, N. (2017a) *Platform Capitalism*. Cambridge: Polity.

Srnicek, N. (2017b) 'We Need to Nationalise Google, Facebook and Amazon. Here's Why'. *Guardian* [online] available from https://www.theguardian.com/commentisfree/2017/aug/30/nationalise-google-facebook-amazon-data-monopoly-platform-public-interest [30 August 2017].

Stahl, M. (2003) 'To Hell with Heteronomy: Liberalism, Rule-Making, and the Pursuit of "Community" in an Urban Rock Scene'. *Journal of Popular Music Studies* 15 (2): 140–65.

Stahl, M. (2013) *Unfree Masters: Recording Artists and the Politics of Work*. Durham, NC: Duke University Press.

Stirling, E. (2016) '"I'm Always on Facebook!": Exploring Facebook as a Mainstream Research Tool and Ethnographic Site'. In *Digital Methods for Social Science: An Interdisciplinary Guide to Research Innovation*, edited by Roberts, S., Snee, H., Hine, C., Morey, Y. and Watson, H. London: Palgrave Macmillan UK, 51–66.

Straw, W. (1990) 'Systems of Articulation, Logics of Change: Communities and Scenes in Popular Music'. *Cultural Studies* 5 (3): 368–88.

Straw, W. (2001) 'Scenes and Sensibilities'. *Public* 22 (23): 245–57.

Street, J. and Phillips, T. (2016) 'What Do Musicians Talk about When They Talk about Copyright?' *Popular Music and Society* 40 (4): 1–12.

Stutzman, F., Gross, G. and Acquisti, A. (2012) 'Silent Listeners: The Evolution of Privacy and Disclosure on Facebook'. *Journal of Privacy and Confidentiality* 4 (2): 7–41.

Sunstein, C. R. (2001) 'The Daily We: "Is the Internet Really a Blessing for Democracy?"' *Boston Review* 26 (3): 4–9.

Sunstein, C. R. (2004) 'Democracy and Filtering'. *Communications of the ACM* 47 (12): 57–59.

Taylor, A. (2014) *The People's Platform: Taking Back Power and Culture in the Digital Age*. Toronto: Random House Canada.

Terranova, T. (2000) 'Free Labor: Producing Culture for the Digital Economy'. *Social Text* 18 (2): 33–58.

Thaler, R. H. and Sunstein, C. R. (2009) *Nudge: Improving Decisions about Health, Wealth, and Happiness*. London: Penguin Books.

Théberge, Paul. (1997) *Any Sound You Can Imagine: Making Music/Consuming Technology*. Middletown, CT: Wesleyan University Press.

Thompson, M. (2017) 'The Discomfort of Safety'. *Society + Space* [online]. https://www.societyandspace.org/articles/the-discomfort-of-safety [12 September 2018].

Thompson, S. (2004) *Punk Productions: Unfinished Business*. Albany, NY: SUNY Press.

Thoreau, H. D. (1986) *Walden and Civil Disobedience*, edited by Meyer, M. Harmondsworth: Penguin.

Till, R. (2016) 'Singer-Songwriter Authenticity, the Unconscious and Emotions (Feat. Adele's "Someone Like You")'. In *The Cambridge Companion to the Singer-Songwriter*, edited by Williams, K. Cambridge Companions to Music. Cambridge: Cambridge University Press, 291–304.

Toynbee, J. (2000) *Making Popular Music: Musicians, Creativity and Institutions*. London: Arnold.

Toynbee, J. (2001) *Creating Problems: Social Authorship, Copyright and the Production of Culture*. Milton Keynes: The Pavis Centre for Social and Cultural Research.

Travers, R. (2017) 'University "Safe Spaces" Are a Dangerous Fallacy – They Do Not Exist in the Real World'. *Telegraph* [online] available from http://www.telegraph.co.uk/education/2017/02/13/university-safe-spaces-dangerous-fallacy-do-not-exist-real/ [27 February 2017].

Treanor, B. (2008) *Slow University: A Manifesto* [online] available from http://faculty.lmu.edu/briantreanor/slow-university-a-manifesto/ [1 September 2017].

Turino, T. (2008) *Music as Social Life: The Politics of Participation*. University of Chicago Press: London.

Turkle, S. (1995) *Life on the Screen: Identity in the Age of the Internet*. London: Weidenfeld & Nicolson.

Turner, F. (2010) *From Counterculture to Cyberculture: Stewart Brand, the Whole Earth Network, and the Rise of Digital Utopianism*. Chicago: University of Chicago Press.

Turner, F. (2017) *Don't Be Evil: Fred Turner on Utopias, Frontiers, and Brogrammers* [online] available from https://logicmag.io/justice/fred-turner-dont-be-evil/ [18 February 2020].

Vaidhyanathan, S. (2004) 'The State of Copyright Activism'. *First Monday* 9 (4/5).

van Alstyne, M. and Brynjolfsson, E. (2005) 'Global Village or Cyber-Balkans? Modeling and Measuring the Integration of Electronic Communities'. *Management Science* 51 (6): 851–68.

Van Dijck, J. (2009) 'Users Like You? Theorizing Agency in User-Generated Content'. *Media Culture & Society* 31 (1): 41–58.

van Dijck, J. (2013) *The Culture of Connectivity : A Critical History of Social Media*. Oxford: Oxford University Press.

Verbuč, D. (2018) 'Theory and Ethnography of Affective Participation at DIY Shows in U.S'. *Journal of Popular Music Studies* 30 (1–2): 79–108.

Virno, P. (2004) *A Grammar of the Multitude*. London: Semiotext(e).

White, E. (1992) 'Revolution Girl-Style Now! : Notes from the Teenage Feminist Rock "n" Roll Underground'. *LA Weekly*, July.

Whitson, J. R. (2013) 'Gaming the Quantified Self'. *Surveillance & Society* 11 (1–2): 163–76.

Withers, D. (2010) 'Transgender and Feminist Alliances in Contemporary U.K. Feminist Politics'. *Feminist Studies* 36 (3): 691–7.

X (1980) 'The Unheard Music'. *Los Angeles* [LP]. Slash, SR-104.

XTC (1978) *Go 2* [LP]. Virgin, V2108.

Zimmerman, A. G. and Ybarra, G. J. (2016) 'Online Aggression: The Influences of Anonymity and Social Modeling'. *Psychology of Popular Media Culture* 5 (2): 181–93.

INDEX

Abercrombie, Nicholas, 'Incorporation/ Resistance Paradigm' 38
Adorno, Theodor W. 104
advertisement/advertisers 4, 7, 51, 59–60, 62, 65, 83, 101, 106, 112, 115, 125, 128, 130, 133, 135
aesthetics 2–4, 6, 30, 34, 37, 47–8, 55, 69, 94
   handmade/handwritten 49–50
   intimate 51
   non-professional 50
   platform 50–1
Albini, Steve 32
alienation 37, 48 n.2, 60–1, 63, 66
Alphabet 15
alternative distribution network 5, 138–48
alternative music 2–6, 23, 70, 113, 131, 137
amateur music 8, 32, 36, 115, 123, 125, 138
Amazon 126 n.5
Android operating system 139
anonymity, online 16 n.4, 68, 80–2, 135
anti-commodification 36 n.2
anti-consumerism 30
anti-racism 13
Apple Music 14, 127
app walkthrough method 106
Arnold, Gina 33
   Route 666: The Road to Nirvana 31
artistic critique 8
Arvidsson, Adam 56
Attitude Is Everything disability-led charity 123 n.1

audience 2, 4–5, 8–10, 15 n.3, 30–2, 35–6, 38, 55, 57–9, 73
   artist–audience 8, 36, 55, 67, 70, 77
   commodity 59–65
   imagined/imagining 105, 108–9
   online 109, 115
   wrong 86–7
audit culture 103, 119
authenticity 12, 27–9, 32, 40, 47, 101, 114, 123 n.1
   Frith on 47 n.1
   intimacy/intimate 47–53
autonomous production 54
autonomy, artistic 3, 7, 17–18, 20, 28, 39–40, 42, 48 n.3, 66, 89, 126, 135
avant-garde music 36 n.2
Azerrad, Michael 33
   Our Band Could be Your Life 31

Baker, Sarah 39
Bandcamp 14, 96, 108–9, 125, 131–2, 142
Banet-Weiser, Sarah 51
Bay Area hip-hop scene 26
Baym, Nancy 8–10, 14, 108–11
Behavioural Insights Limited (Behavioural Insights Team) 131 n.8
Benjamin, Walter 36 n.2, 49
Biafra, Jello 2
Bikini Kill zine 34, 86
Black Flag band 31
Blauner, Robert 60–1, 63
Boehringer, Jorge 55
Boltanski, Luc 9
   The New Spirit of Capitalism 8
Bourdieu, Pierre 4, 54

INDEX

Bowlie (Anorak) indie-pop forum 78. *See also* music forums
boyd, danah 74, 108, 118
brass band movement 72
Braverman, Harry 20, 122, 127
   Labour and Monopoly Capital 124
   scientific management 124, 126
Breeze, Maddie 16 n.4
Bristol, research project in 26
Brown, Wendy 42, 46
Brudenell Social Club 12–13
built environments, online 9–10

Campaign Choirs Writing Collective (2018) 36
capitalism 8, 41, 60–2, 66, 122, 126, 146
   communicative 118, 144, 147
   democratic 139
   Fordist 124
   platform 8, 68, 125, 127, 131–2, 134, 137, 140, 146
censorship 27 n.1, 52
Chiapello, Eve 9
   The New Spirit of Capitalism 8
Cohen, Julie E. 116 n.2
collectivism 4, 19, 39, 68–77, 79, 82
commercial/commercialism 2, 4–6, 8, 19–20, 28, 32–3, 35, 133
commodity/commodification 7–8, 29–30, 32, 34, 36–8, 47–8, 61, 83–4, 83 n.1, 104, 113
   audience 59–65
   cultural 127
   handmade/hand-replicated 48
   Marx on 48 n.2, 49
   self-commodification 118–19
'commons'-style internet 68–9
communication 46–7, 51, 58, 61, 95, 97, 128, 133
   artist–audience 70
   authentic 8, 52
   intimacy/intimate 8, 48, 114
      epistolary intimacy 47, 51, 104
   mass 31, 34, 51, 115, 142
   online 3, 9–11, 19, 50, 53, 87, 98
competition, music 2, 8, 35, 107, 112, 128, 133, 135–7

convivial competition 68–77
   online 81
consumer/consumerism 20, 30, 32, 37, 59–60, 66
consumption 21, 24, 27, 37, 40, 60, 65–6, 136
contingency/contingent 127, 127 n.7
copyright 7, 76, 82, 83 n.1, 116 n.2, 143
creativity, musical 19, 25, 27, 76, 82, 116 n.2, 127 n.7
Criminal Justice and Public Order Act (1994) 25
CTEA (Copyright Term Extension Act) 143
cultural activity 23–5, 39, 42, 125
cultural gatekeepers 19, 68, 70–1
cultural industries 5, 17, 41, 54, 83, 122, 124–5, 126 n.5, 127
cultural production 3, 8, 27, 30, 37–8, 49, 52, 66, 122, 128, 136
cultural resistance 6, 8, 13, 18–20, 24, 37–43, 67, 71, 86

datafication 104, 119
data mining tools, social media 111
Decolonise Fest, 2017 58
deconstructive punk 30, 32
democratization, cultural 1–3, 27, 49–50, 87
The Desperate Bicycles band 30
digital divide 27, 87
digital music 14, 27, 32
digital production 122
DistroKid music service 125
diversity of music 23–4, 26, 28–9, 58
DIY Access Guide (2017) 123 n.1
DIY Diaspora Punx 58
DIY (do-it-yourself) scene 1, 9–10, 28, 74
   case study 97–101
   friction 92–7
   Glasgow DIY scene 39
   historical 28–35, 49
      riot grrrl scene (1989–96) 6 n.1, 10, 24, 29, 33–5, 40, 86–7, 97, 104
      UK post-punk scene (1978–83) 24, 29–31

US post-hardcore indie scene
        (1983–8) 24, 29, 31–3
    insularity and openness 19, 33,
        37–8, 86–8, 93, 96–7, 116
    musical activity 6–7, 9, 17, 23–8,
        46, 59, 65, 97, 118
        and COVID-19 pandemic 88 n.1
    ownership 19, 60, 68, 76–82
    promoters/promotion 15 n.3, 16–
        17, 45, 57, 59, 62, 64, 69–70,
        79, 91, 97–8, 128–9, 135
        self-promotion 8, 79, 82, 128,
            145
    research in Leeds 11–18, 23
        interviews of practitioners
            15–16, 45, 47, 52–3, 61–2, 62
            n.7, 64, 67, 73–5, 98, 100–1,
            145
        music venues 12–13
        racism 13
        research and practice 16–18
        social media platforms 14–15
    skill-sharing 50
    togetherness 67, 84
    UK DIY scene 5, 27
DIY Space For London (DSFL), 2016
    57, 70–1, 92 n.3
Downes, Julia 16 n.4, 33, 40
Drake 115
Dropbox 125
Duncombe, Stephen 39–40
Dyer-Witheford, Nick 61, 68

economy 3, 25, 56 n.5, 60, 125, 147
    attention 73, 80, 129–30
    ethical 56
    moral 146
    online 56
    platform 20, 135, 138
    political 42, 128
electronic music, music genre 1, 12,
    14, 29
Ello, social network 64
emotional intimacy 34, 51
empowerment 3, 40, 42, 46–7, 63, 66
Entertainment! (1979), Gang of
    Four 30
environment/environmentalism 10, 148

epistolary intimacy 47, 51, 104
ethics, DIY 7–10, 15, 16 n.4, 26, 42–3,
    47, 57, 68, 70, 79, 83, 113, 118,
    124, 134–7, 141

Facebook 10, 14–16, 59–65, 97–9,
    103, 116, 123, 127, 129, 131,
    137, 139, 141, 143–4
    enterprise discourse 20, 79, 105,
        107, 109, 118
    Facebook Events 91, 97–8, 125
    Facebook Group 16, 68, 78–9, 106
    Facebook Pages 14, 16, 50, 68,
        79–80, 82–3, 97, 101, 105–8,
        112, 125
        administrators of 106–7
        Likes 106–7, 109–13, 117, 134,
            146
        Sponsored Posts 112–13
    guidance 130
    Messenger 15, 98, 125
    real names policy 65
fanzine 28, 38, 77
feminism/feminist 6, 12, 16, 33–4,
    41–2, 58, 74
    safe spaces 89
Finn, Ed 51
Fisher, Mark 4, 146
FLOSS (Free/Libre Open-Source
    Software) 139, 143
folk music genre 12, 16, 41, 72, 148
FOMO (fear of missing out) 146
Fordist model 122, 124–5, 136
framing of social media 143
Fraser, Nancy 24, 41–3, 59, 70, 149
    misrecognition/maldistribution
        41–2
    participatory parity 71
    recognition 90
'free lunch' model 7, 60
Frith, Simon 11
    on authenticity 47 n.1

Geffen, Sasha 58 n.6
Giddens, Anthony 56
Glasgow DIY scene 39
Golpushnezhad, Elham (study of rap
    music in Iran) 26–7

Google 14–15, 91, 126 n.5, 137, 139
Go 2 (1978), XTC 30
Griffin, Naomi 16 n.4
Grosser, Benjamin 103, 110
guitar-based music 114, 116
Gurak, Laura 108

Halpin, Harry 143
hardcore punk 12, 31–2
Harrison, Anthony Kwame 27–8, 38
Harvie, David, convivial competition 68–76
Hayler, Rob, 'no-audience underground' 15 n.3, 54–5, 54 n.4
headline act 71
Hearn, Alison 51
Herstand, Ari, *How to Make It in the New Music Business* 1
Hesmondhalgh, David 8, 39
   Why Music Matters 19 n.5
heteronomous production 54
hierarchy 19, 68–9, 78–9, 81, 83, 90
   anti-hierarchical 67–8
   artist–audience 55, 67
   hierarchization 69
   pop 35
   Stahl on 70
Hui, Yuk 143

ICT (information and communication technologies) industries 3, 7–8, 18
immaterial labour 61, 110, 147
indie/independent music genre 1, 4, 7–8, 14, 31, 70, 72
   in Philippines 27
   US post-hardcore indie scene (1983–8) 24, 29, 31–3
indie-punk 5, 12, 23, 35, 50, 58, 69, 138, 148
Instagram 15, 80
intimacy/intimate authenticity 47–53
Iran, study of rap music in 26–7
Irish music fanzine culture, study of 28, 38
iTunes (Apple Music) 14

Jurgenson, Nathan 118–19

Kahn-Harris, Keith 11, 95
Kanai, Akane 58–9
Kennedy, Helen 111
Kinder, Kimberley, *DIY Detroit* 25

labour process 20, 122, 124
Lamm, Nomy 34, 51
LANDR service 126 n.3
Langlois, Ganaele 142, 146
liberation 55
libertarian paternalism 130–1
'lo-fi' (bedroom) technique 49
Lohman, Kirsty 12 n.2
Longhurst, Brian, 'Incorporation/Resistance Paradigm' 38

MacKaye, Ian 32
   'Guilty of being white' 41 n.5
mainstream music 2, 4, 6–7, 18, 26, 36, 46, 53, 79, 82, 94, 101, 105, 114
marginality 41, 41 n.4
marketing 112, 128, 130, 132–5
   anti-marketing 132–3, 135
   and branding 8, 122
   Facebook and 65, 125
   online 113
Marx, Karl 48 n.2, 49, 60, 63, 113, 124–5
   Capital, Vol. 1 126 n.6
mass communication 31, 34, 51, 115, 142
mass media 36, 60
Mattern, Mark, *Acting in Concert. Music, Community And Political Action* 96
McKay, George, *DiY Culture: Party and Protest in Nineties Britain* 25
metrics, social media 103, 109–13, 115–17, 119
   accuracy/inaccuracy 109–10
   and context collapse 108–9
   failure of 109
   and platform discourse 105–8
   quantitative 109–10, 119

Miller, Daniel 66
The Minutemen band 32
music criticism 4, 114
music culture 1, 4, 6, 11, 16–17, 25–6,
  28–9, 35, 37, 45, 66, 94, 138,
  149
  of black youth in UK 27 n.1
  popular 1, 6, 29, 35, 148–9
  study of 28
Music for Misfits (television program)
  41 n.4
music forums 15, 19, 68, 77–8
music industry 1–3, 5–7, 12, 17–18,
  26, 29, 35–6, 46, 48, 48 n.3
  mental health within 64, 145–7
music-making 5, 11, 29, 49, 76, 80, 83,
  114, 145, 147
music streaming services 14. See also
  specific services
MySpace 15, 50, 68, 77

networked publics approach 74–5
non-professional music 8, 123
not-for-profit approach 5, 7, 69–71
nudge theory 130–3

omniopticon model of social media
  118–19
online activity surveillance 7, 56
Open Data, UK government's 139
open-source software 20, 138–43
optimization 20, 122, 139
  and social media 127–36

Pearce, Ruth 12 n.2
peer-production 68, 139, 143
perzines (personal zines) 34
Philippines, indie in 27
physical *vs.* ideological neutralization 9
piracy 68, 143
platform (social media) 9, 11, 14,
  18–20, 42, 47, 50–1, 53, 56–7,
  58 n.6, 66, 74, 83, 111, 115, 136
  aesthetic 50–1
  affordances 10, 111, 118
  alternative distribution 138–48
  bridging/bonding 100
  DIY and 148–9
  and deskilling 123–7
  imagination 113–16
  echo chambers 19, 87–8, 97
  Facebook (*see* Facebook)
  functionalities/tools 91, 103
  glamour 51
  Instagram 15, 80
  metrics and (*see* metrics, social
    media)
  music streaming services 14 (*see
    also specific services*)
  MySpace 15, 50, 68, 77
  optimization and 127–36
  ownership on 19, 60, 68, 76–82
  photo and video-centric 15
  platform capitalism 8, 68, 125, 127,
    131–2, 134, 137, 140, 146
  platform-dominated internet 3
  relatability 19, 47, 56–9
  representation 53–9, 61, 111–12,
    118, 135
  self-branding 19, 57, 80, 128,
    136, 146
  and social authorship 82–4
  strategies of using 144–8
  Twitter 10, 14, 52, 56, 130
  and users/consumers 3, 9–10, 59–65
  'Web 2.0' platforms 68, 142
  YouTube 3, 14–15, 21, 87, 125,
    139–41
platformization 122–3, 127–8, 136
politics 9–10, 16, 25, 40–1
  feminist and queer 6, 12, 16
  French radical politics of 1960s 8
  racial 6
  representational 5, 20
pop/popular culture 2, 5, 30, 52–4, 94
poptimism 4, 114
popularity 71, 85, 105, 117, 119, 121
popular music 1, 3–4, 6, 8–9, 12, 19,
  28–9, 46–7, 53, 67–8, 70, 85,
  103, 125, 127, 138, 148
  and community-based political
    action 96
  core units of 80
  and DIY imagination 113–16
  and DIY music 24, 35–8, 87,
    92, 105

INDEX

post-punk music genre 1, 12, 29–31, 53, 77, 90, 101, 104
practitioners, DIY 1, 5–8, 12 n.2, 13–15, 15 n.3, 16 n.4, 18–20, 24, 28–9, 31–2, 36–9, 41 n.5, 46, 49, 52, 91, 138
  commodities 48
  'firsts' (band/show/song) 45
  friction 92–7
  hip-hop 38
  indie-punk 12
  interviews of 15–16, 26, 45, 47, 52–3, 61–2, 62 n.7, 64, 67, 73–5, 98, 100–1, 123, 140, 145
  liberation 55
  ownership 77
  participation 67, 73
  post-punk 30
  publics 85–6, 92
  punk 39
  rap 26–7
  safe/safer space 89–93, 89 n.2, 100
  as social media user 59
  subcultural 40
pub rock band 98–9, 101
punk music genre 1–2, 12–13, 18, 30–2, 49, 53–4, 58, 86

quantification 103–4, 118
queer 6, 12–13, 12 n.2, 16, 91, 97, 101, 135

race/racism 2, 5–6, 6 n.1, 13, 40, 99
Radio Free Midwich blog 54
rave music 1, 10
Real Name policies 65, 115
record labels 5, 7, 31, 48 n.3, 52, 76, 125
reification process 25, 104, 110–11, 116, 118–19
remix culture 27, 68
Resonate music streaming platform 139
Reynolds, Simon 54
Rey, PJ, 'ambient production' 63
Riot grrrl 51, 54, 76, 86–7, 97, 104
  DIY Diaspora Punx 58
  scene (1989–96) 6 n.1, 10, 24, 29, 33–5, 40

rock music 12, 29, 35, 70
rockism 114
Rough Trade Records 30
Ryan, Ciarán (study of Irish music fanzine culture) 28, 38

safe/safer space policy 89–93, 89 n.2, 101
Saga platform 141
scepticism 4, 27, 138
Scritti Politti band 77, 122, 124
  Cupid and Psyche 85 126 n.4
  on *Grapevine* show 121
  'Messthetics' 30
  musical 'heyday' 126, 126 n.4
self-expression 48, 51, 54–9, 61, 63, 130, 141, 147, 149
self-governance 42–3, 108
self-realization 2, 39–40, 45–6, 54, 56, 59, 66, 80, 115, 144
Sheeran, Ed 48
signal boosting 74
Silicon Valley 14, 75
simulacrum 111
Slow University movement 144–5
Smythe, Dallas 60
Snapchat 15
social authorship 82–4, 83 n.1
social critique 8–9
social justice 3, 6, 9, 13, 24, 41–2, 52, 68, 71, 74, 136, 148–9
social media. *See* platform (social media)
social movement 6, 24, 26, 35, 39
SoundCloud 14, 81, 125, 141
  autoplay 74
spam/spamming 78–9
Spotify 14, 21, 80, 127
Srnicek, Nick 125, 148
Stahl, Matt 69–70
  reciprocity 72
status order 68, 70–1
straight edge punk scene (Brazilian), research on 28
subcultural activity 40, 50, 54, 122
subjectivation 146
subjectivity/subjectivities 42, 47, 54, 60, 94, 116 n.2, 118, 143–4, 146–7
Sunstein, Cass 130, 131 n.8

Taylor, Frederick W. 124, 131
Taylorism/Taylorist model 122, 124–5
Teen Idles band 32
Tennant, Neil 114
Thaler, Richard 130, 131 n.8
third-party distributors 14, 17, 62, 135
Thompson, Marie 90
Toynbee, Jason 82–3
transformative process 45–6
transgression 11, 95–6
trans music genre 12 n.2, 58 n.6
trigger warnings 89
Tunecore 14, 125
Twitter 10, 14, 52, 56, 130

UK Conservative–Liberal Democrat coalition government 131 n.8
UK DIY Group 79
UK grime scene 27
UK post-punk scene (1978–83) 24, 29–31
underground music 4, 6
  in Iran 26

Usenet discussion groups 87–8
US post-hardcore indie scene (1983–8) 24, 29, 31–3

van Dijck, Jose 59, 104
visibility/invisibility in DIY 5, 47, 56, 80, 91, 114, 128–9

Weber, Max, spirit of capitalism 8
'Web 2.0' platforms 68, 142
Wharf Chambers 12–13, 90, 92 n.3, 93, 95, 97, 115
WhatsApp application 15, 125
Wikipedia 139
Wire band 31
workplace alienation 60

YouTube/YouTuber 3, 14–15, 21, 87, 125, 139–41

zines 8, 31, 33–4, 45, 49–51, 76, 86
Zoombombing 88 n.1
Zuckerberg, Mark 112

www.ingramcontent.com/pod-product-compliance
Ingram Content Group UK Ltd.
Pitfield, Milton Keynes, MK11 3LW, UK
UKHW020856170225
455196UK00008B/337